Practical Guide to

Oracle SQL,
T-SQL and MySQL

Practical Guide to
Oracle SQL,
T-SQL and MySQL

Preston Zhang

CRC Press
Taylor & Francis Group
Boca Raton London New York

CRC Press is an imprint of the
Taylor & Francis Group, an **informa** business

CRC Press
Taylor & Francis Group
6000 Broken Sound Parkway NW, Suite 300
Boca Raton, FL 33487-2742

First issued in paperback 2018

ISBN-13: 978-1-138-10518-8 (hbk)
ISBN-13: 978-1-138-34752-6 (pbk)

Library of Congress Cataloging-in-Publication Data

Names: Zhang, Preston, author.
Title: Practical guide to Oracle SQL, T-SQL and MySQL / Preston Zhang,
    database administrator, University of Georgia, Watkinsville, Georgia, USA.
Description: Boca Raton : CRC Press, [2017] | "A science publishers book." |
    Includes bibliographical references and index.
Identifiers: LCCN 2017040116 | ISBN 9781138105188 (hardback : alk. paper)
Subjects: LCSH: SQL (Computer program language) | Oracle (Computer file)
Classification: LCC QA76.73.S67 Z54 2017 | DDC 005.75/6--dc23
LC record available at https://lccn.loc.gov/2017040116

Visit the Taylor & Francis Web site at
http://www.taylorandfrancis.com

and the CRC Press Web site at
http://www.crcpress.com

Preface

Databases are used everywhere. They effect on our daily lives widely. Online business companies use databases to store critical data for their products and users; Doctor offices use databases to keep patient, pharmacy and insurance information; Banks use databases to track millions of financial transactions.

Relational database management systems (RDBMS) have become the standard database type from 1980s. The most popular relational database management systems in the world are Oracle, SQL Server and MySQL. To get data or manipulate data from database systems developers and database administrators use Structured Query Language (SQL).

I have worked on Web applications using Oracle and MySQL databases on the backend. In order to display important data I write SQL statements to access databases in php or other development tools. I also use SQL to create databases or update database structures. SQL is so powerful that I can process millions of records in few seconds.

As a database administrator I have been working on Oracle, SQL Server and MySQL for decades. Although the basic SQL statements for Oracle SQL, SQL Server T-SQL and MySQL are similar to each other, some functions and styles are quite different. I often need to work with different database systems at the same time and it takes time for me to check SQL syntax for the three database systems. There are a lot of SQL books available in the market, but it is very hard to find a practical SQL book that comparing the differences between the three major database systems. That's why I want to write this reference book with step by step examples in the real working environment.

I hope that this book can be a quick reference book for Oracle SQL, SQL Server T-SQL and MySQL.

Why Learn SQL?

- SQL is one of the most desirable programming skills
- SQL is used by all types of career fields
- You can use SQL to ask questions about your business
- You can get useful business reports from SQL statements
- You can manipulate millions of records in seconds
- You can import data to a database
- You can export data from a database
- You can embed SQL statements to other programming languages

Who This Book Is For

This book is for beginning and intermediate SQL developers, database administrators, database programmers and students. It starts from database concepts, installation of database management systems, database creation and datatypes. It introduces basic and advanced SQL syntax with side by side examples in Oracle SQL, T-SQL and MySQL. The SQL code in this book is fully tested in Oracle 12c, SQL Server 2012 and MySQL 5.7.

How to Use This Book

To run the examples from this book you need to install the following database systems and development tools:

> Oracle 11g or 12c
> Oracle SQL Developer
> SQL Server 2012 or above
> SQL Server Management Studio 2012 or above
> MySQL Server 5.7
> MySQL Workbench 6.3

All of the above software can be download from Oracle.com and Microsoft.com

Acknowledge

I wish to express appreciation to the Science Publisher editors who have been supporting this book from the beginning and made this book a reality.

My deepest expression of gratefulness goes to my mom who has been learning for 30 years after her retirement.

Contents

Preface   v

Chapter 1   Introduction to SQL and Relational Databases   1

Brief History of SQL and Relational Databases   2
SQL Standards   2
Oracle, SQL Server and MySQL Versions   3
Relational Database Basic Concepts   3
Constraints   7
Data Integrity   7
Types of Relationships   8
    One-to-Many Relationships   8
    Many-to-Many Relationships   9
    One-to-One Relationships   9
    Self-Referencing Relationships   9
Summary   10

Chapter 2   Data Types   11

Character Data Types   11
Number Data Types   12
Date and Time Data Types   13
Boolean Data Type   14
Summary   14

Chapter 3   Installation of Oracle, SQL Server and MySQL   15

Minimum System Requirements   15
Installation of Oracle 12c   16
Installation of SQL Server 2016   20
Installation of MySQL Server 5.7   26
Summary   32
Exercise   33

Chapter 4   Database Development Tools   34

Command Line Tools   34
    Oracle SQL Plus   34
    MySQL Command Line Client   36
Installation of Oracle SQL Developer 4.3   38
Installation of SQL Management Studio 2016   40
Installation of MySQL Workbench 6.3   43
Summary   46
Exercise   46

Chapter 5	**Data Definition Language (DDL)**		**47**
	Data Definition Language Statements		47
		Using SQL Commands to Create a Database	48
		Using GUI Tools to Create a Database	49
		Using SQL Commands to Create a Table	51
		Using GUI Tools to Create a Table	54
		Using Data from an Existing Table to Create a Table	56
		Renaming a Table	58
		Truncating a Table	61
		Altering a Table	61
	Summary		64
	Exercises		64
Chapter 6	**Data Manipulation Language (DML)**		**65**
	Data Manipulation Language Statements		65
		INSERT INTO Statement	65
		SELECT Statements	72
		DISTINCT Clause	73
		WHERE Clause	74
		Arithmetic Operators	74
		Order of Arithmetic Operators	76
		Comparison Operators	76
		AND Condition	77
		OR Condition	77
		IN Condition	78
		BETWEEN Condition	79
		IS NULL Condition	79
		IS NOT NULL Condition	80
		LIKE Condition	81
		ORDER BY Clause	82
		ALIASES	84
		INSERT Multiple Records into an Existing Table	85
		UPDATE Statement	86
		DELETE Statement	86
	Data Control Language		87
	Summary		87
	Exercises		87
Chapter 7	**Aggregate Functions and GROUP BY Clause**		**89**
	Aggregate Functions		89
		AVG ( )	90
		COUNT ( )	90
		MIN ( )	91
		MAX ( )	92
		SUM ( )	93
	GROUP BY and HAVING Clause		94
	GOUNP BY with AVG ( ) Function		94
	GROUP BY with COUNT ( ) Function		95

	GROUP BY with HAVING Example	96
	Summary	97
	Exercises	97
Chapter 8	**Functions**	**98**
	Common Number Functions	98
	CEIL ( )	98
	CEILING ( )	98
	FLOOR ( )	99
	GREATEST ( )	99
	LEAST ( )	100
	MOD ( )	100
	POWER ( )	101
	ROUND ( )	101
	SQRT ( )	102
	TRUNC ( )	102
	Common String Functions	103
	CONCAT ( )	104
	FORMAT ( )	106
	LEFT ( )	106
	INITCAP ( )	106
	LENGTH ( )	107
	LEN ( )	107
	LOWER ( )	108
	LPAD ( )	108
	LTRIM ( )	109
	REPLACE ( )	109
	RIGHT ( )	110
	RPAD ( )	110
	RTRIM ( )	111
	SUBSTR ( )	111
	SUBSTRING ( )	111
	UPPER ( )	112
	Common Date and time Functions	113
	CURRENT_TIMESTAMP	113
	ADD_MONTHS ( )	114
	DATEADD ( )	114
	DATE_ADD ( )	114
	EXTRACT ( )	114
	DATEPART ( )	114
	CURRENT_DATE	115
	GETDATE ( )	115
	CURRENT_DATE ( )	115
	MONTHS_BETWEEN ( )	115
	DATEDIFF ( )	116
	PERIOD_DIFF ( )	116
	SYSDATE	116
	SYSDATETIME ( )	116
	SYSDATE ( )	116

	Conversion Functions	116
	CAST ( )	117
	TO_DATE	117
	CONVERT ( )	118
	STR_TO_DATE ( )	119
	Summary	120
	Exercises	121
Chapter 9	**Advanced SQL**	**122**
	Advanced SQL Statements	123
	Union	123
	Union All	124
	INTERSECT	124
	EXCEPT	125
	MINUS	125
	ROWNUM	126
	TOP	126
	LIMIT	126
	Subquery	128
	CASE	130
	SEQUENCE	132
	IDENTITY	132
	AUTO_INCREMENT	132
	Summary	136
	Exercises	136
Chapter 10	**Joins**	**137**
	INNER JOIN	137
	JOIN with USING Clause	138
	Joining with Multiple Tables	141
	LEFT JOIN	142
	RIGHT JOIN	142
	FULL JOIN	142
	Summary	144
	Exercise	145
Chapter 11	**Views**	**146**
	Creating Views in Oracle	146
	Creating Views in T-SQL	149
	Creating Views in MySQL	151
	Updating Views	153
	Summary	154
	Exercise	154
Chapter 12	**Data Import and Export**	**155**
	Oracle Data Export from Query Results	155
	SQL Server Data Export from Query Results	157
	MySQL Data Export from Query Results	159
	Oracle Data Import Tool	161

SQL Server Data Import Tool   166
MySQL Data Import Tool   171
Summary   174
Exercise   174

Chapter 13   Stored Procedures   **175**

Steps to Create an Oracle Stored Procedure   175
Steps to Create a SQL Server Stored Procedure   176
Steps to Create a MySQL Stored Procedure   177
A Stored Procedure with Parameters   178
Summary   182
Exercise   183

Index   **185**
About the Author   **189**

Chapter 1

Introduction to SQL and Relational Databases

Relational database management systems (RDBMS) have become the standard database type for various industries since the 1980s. These systems allow the users to store data and access data in graphic user interfaces. It also allows users to set security rules.

Structured Query Language (SQL) is a standard computer language for relational database management systems. SQL has different dialects. For example, Oracle SQL is called PL/SQL, MS SQL Server SQL is called T-SQL (Transact-SQL).

SQL is a very useful tool for database developers and database administrators. Database developers use SQL to select, insert, and update data. Database administrators (DBAs) apply their SQL skills to support Oracle, SQL Server, MySQL and other database systems.

The highlights of this chapter include

- Brief History of SQL and Database Systems
- SQL Standards
- Oracle, SQL Server and MySQL Versions
- Introduction to RDBMSs
- Relational Database Basic Concepts
- Entity Relational Diagram Used in This Book

Brief History of SQL and Database Systems

Table 1.1 History of SQL and Database Systems

Year	SQL and Database Development
1970 to 1972	Dr. E.F. Codd in IBM introduced in his paper the term "A Relational Model of Data for Large Shared Data Banks". In the paper he defined RDBMs by Codd's 12 rules.
1970s	Ingres and System R were created at IBM San Jose. System R used the SEQUEL query language. The development of SQL/DS, DB2, and Oracle were based on the SEQUEL query language.
1976	Dr. Peter Chen developed the entity-relationship model. This model becomes the foundation of many systems analysis and design methods.
1980s	Structured Query Language became the standard query language. Computer sales increased rapidly. Relational database systems became a commercial success. IBM's DB2 and IBM PC resulted in the launches of many new developments of database systems such as PARADOX, dBase III and IV.
1990s	Successful Online businesses let to demand for database accessing tools. MySQL and Apache became open source solution for the Internet. Application development tools including Oracle Developer, Power Builder, and Visual Basic were released.
2000s	The three leading relational database systems in the world are Oracle, Microsoft SQL Server and MySQL.

SQL Standards

Table 1.2 SQL Standards

Year	SQL Standard
1974	Original SQL (SEQUEL)
1986	SQL became a standard by ANSI (American National Standards Institute) and ISO (International Standards Organization)
SQL/96	Major modification (ISO 9075)
SQL/99	Added many features including recursive queries, triggers, procedural and control-of-flow statements, and some object-oriented structures
SQL/2003	Introduced XML-related features
SQL/2006	Defined ways for importing and storing XML data in database
SQL/2008	Added TRUNCATE TABLE statement and INSTEAD OF triggers

Oracle, SQL Server and MySQL Versions

Table 1.3 Different versions for the three database systems

Oracle	SQL Server	MySQL
1979–Oracle 2	1989–SQL Server 1.0	1995–First Release
1983–Oracle 3	1991–SQL Server 1.1	1996–MySQL 3.19
1984–Oracle 4	1993–SQL Server 4.21	1997–MySQL 3.20
1985–Oracle 5	1995–SQL Server 6.0	1998–MySQL 3.21
1988–Oracle 6	1996–SQL Server 6.5	2000–MySQL 3.23
1992–Oracle 7	1998–SQL Server 7.0	2002–MySQL 4.0
1997–Oracle 8	2000–SQL Server 2000	2003–MySQL 4.01
1998–Oracle 8i	2005–SQL Server 2005	2004–MySQL 4.1
2001–Oracle 9i	2008–SQL Server 2008	2005–MySQL 5.0
2003–Oracle 10g	2010–SQL Server 2008 R2	2010–MySQL 5.5
2007–Oracle 11g	2012–SQL Server 2012	2013–MySQL 5.6
2013–Oracle 12C	2014–SQL Server 2014	2015–MySQL 5.7
	2016–SQL Server 2016	2016–MySQL 8.0

Relational Database Basic Concepts

- *Databases*

Relational Database Management System consists of one or more databases.
For example, the following SQL Server has HR and Sample databases.

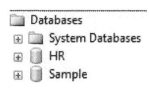

Figure 1.1 Database examples

- ***Entity***

Entity is any person, place, or thing that the data can represent in a database design. For example, Employees and Departments are entities. Entities are converted to tables at the physical design stage.

- **Data Type**

SQL developers need to choose a data type for each column when creating a table. The common data types are boolean, integer, float, currency, string, date and time.

- **DDL**

DDL stands for Data Definition Language. DDL commands can be used to create, modify database structures. Sample DDL commands are CREATE, ALTER and DROP.

- **DML**

DML stands for Data Manipulation Language. DML commands can be used to insert data into database tables, retrieve or modify data, deleting data in database. Sample DML commands are INSERT, DELETE and UPDATE.

- **DCL**

DCL stands for Data Control Language. DCL commands can be used to create rights and permissions. Sample DCL commands are GRANT and REVOKE.

- **Query**

SQL developers can use a query to get data or information from one or more database tables.

Attributes

Entity has its own attributes. For example, an Employee entity may have name, email, phone and salary as attributes.

Types of Attributes

Simple attribute—An attribute that cannot be divided into subparts. For example, an employee's age is a simple attribute.

Composite attribute—An attribute that can be divided into simple attributes. For example, an employee's name has First_Name and Last_Name.

Derived attribute—An attribute whose value can be derived (calculated) from other attribute. For example, Average_Age for employees can be calculated.

Single-value attribute—An attribute contains a single value. For example, City or State.

Multi-value attribute—An attribute have more than one values. For example, an employee can have more than one skills.

Tables

Each database contains collection of tables.

For example, the HR databases has country, customer, departments, employees, job and locations tables.

Figure 1.2 Table examples

- *Fields (Columns)*

Each table consists of smaller entities called fields or columns.

For example, The Country table has three fields (columns): Country_ID, Country_Name and Region_ID.

- ## *Records (Rows)*

Each table consists of one or more records (rows).

For example, the COUNTRY table has the following rows:

COUNTRY_ID	COUNTRY_NAME	REGION_ID
AR	Argentina	2
AU	Australia	3
BE	Belgium	1
BR	Brazil	2
CA	Canada	2
CH	Switzerland	1
CN	China	3
DE	Germany	1

Figure 1.3 Record examples

- ## *Primary Key*

Each table can have only one primary key.

For example, the COUNTRY table has a primary key **COUNTRY_ID.**

- ## *Foreign Key*

Database tables might be related by (foreign key) common column(s).

For example, Location_ID is the common column for Departments and Locations tables.

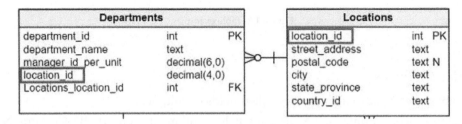

Figure 1.4 Foreign key example

▪ *NULL*

Null value is a field with no value. It is different with a zero value and it has been left blank during record creation.

▪ *Constraints*

Constraints define rules to restrict what values can be stored in columns. This assurances the correctness of the data in the database. For example, we can set a primary key for a table so that there is no duplicated rows in the table.

Common Constraints

- **NOT NULL**—A column does not accept NULL values.
- **DEFAULT**—Set a default value to a column when no value is specified to a column.
- **UNIQUE**—No duplicated values in a column.
- **Primary Key**—A column or a combination of columns that uniquely defines a row. The primary key column can not contain a NULL value.
- **Foreign Key**—A foreign key in one table points to a candidate key in another table.
- **CHECK**—Check whether the value is valid or not.

▪ *Data Integrity*

- **Entity Integrity**—No duplicate records in a table.
- **Referential Integrity**—Referential integrity is violated when deleting a row that is referenced by a foreign key in another table.

For example, a user can't delete the Marketing department from the Departments table, as there are two employees working for the Marketing department (#20). Deleting the Marketing department violates the referential integrity rule. See the sample records below:

DEPT_ID	DEPT_NAME	MANAGER_ID	LOCATION_ID
10	Administration	200	1700
20	Marketing	201	1800

Figure 1.5 Sample data in Departments table

EMPLOYEE_ID	FIRST_NAME	LAST_NAME	EMAIL	PHONE	HIRE_DATE	JOB_ID	SALARY	MANAGER_ID	DEPT_ID
100	Douglas	Grant	DGRANT	650.507.9844	13-JAN-08	SH_CLERK	2600	101	50
101	Adam	Fripp	AFRIPP	650.123.2234	10-APR-15	SH_MGR	8200	109	50
102	Jennifer	Whalen	JWHALEN	515.123.4444	17-SEP-13	AD_ASST	4400	108	10
103	Michael	Hartstein	MHARTSTE	515.123.5555	17-FEB-14	MK_MGR	13000	109	20
104	Pat	Fay	PFAY	603.123.6666	17-AUG-15	MK_REP	6000	103	20

Figure 1.6 Sample records in Employees table

Entity Relational Diagram (ERD) Used in This Book

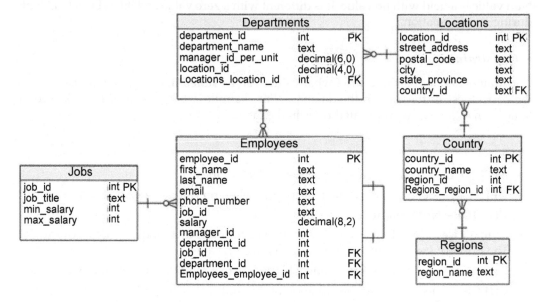

Figure 1.7 Simplified Oracle HR Schema

Types of Relationships

- One-to-Many Relationships
- Many-to-Many Relationships
- One-to-One Relationships
- Self-Referencing Relationships

One-to-Many Relationships

One-to-Many Relationships define the situation when each row in the table_1 has many linked rows in table_2. It is the most common type of relationship.

From the Entity Relationship diagram we can see:

The relationship between the **Employees** and **Departments** is a one-to-many relationship. The Dept_ID is the primary key in the **Departments** table and the foreign key in the **Employees** table. One DEPARTMENT_ID can relate to many rows in the **Employees** table. One department can have one or many employees; an employee is assigned to one department.

The relationship between the **JOB** and **Employees** is a one-to-many relationship. The Job_ID is the primary key in the JOB table and the foreign key in the **Employees** table. One Job_ID can relate to many rows in the **Employees** table. One job title can be used for one or many employees; however, an employee only can have one job title.

The relationship between the **Locations** and **Departments** is a one-to-many relationship. The Location_ID is the primary key in the **Locations** table and the foreign key in the **Departments** table. One Location_ID can relate to many rows in the **Departments** table. One location can have one or many departments; a department only has one location.

The relationship between the **Country** and **Locations** is a one-to-many relationship. The Country_ID is the primary key in the **Country** table and the foreign key in the **Locations** table. One Country_ID can relate to many rows in the **Locations** table. One country can have one or many locations (States or Provinces); a location (State or Province) only belongs to one country.

The relationship between the **Regions** and **Country** is a one-to-many relationship. The Region_ID is the primary key in the **Regions** table and the foreign key in the **Country** table. One Region_ID can relate to many rows in the **Country** table. One region have one or many countries; a country only belongs to one region.

Many-to-Many Relationships

A record in table_1 has many matching records in table_2, and a record in table_2 has many matching records in table_1. For example, an employee may work on one or more projects, and each project may have one or more employees. In this case, MANY employees are related to MANY projects.

How can we build many-to-many relationship in a database system? Suppose we have finished two tables: Employees table and Projects table. We can accomplish many-to-many relationships by creating two one-to-many relationships and adding a link table between the two tables. For example, we can create a table "Emp_Project" that has a composite Primary Key that consists of the two primary keys from the Employees table and Projects tables. Thus, the two one-to-many relationships are:

1. From Employees table to Emp_Project table: One-to-Many relationships.
2. From Project table to Emp_Project table: One-to-Many relationships.

One-to-One Relationships

One-to-Many Relationships define the situation when one row in table_1 has one linked row in table_2.

For example, every person has a social security number. We can create a Person table with name, address, email, phone info and a Person_2 table with social security number. We link the two tables with a key.

Self-Referencing Relationships

A database model with a relationship to itself.

For example, Adam (Employee_ID 101) has a manager (Manager_ID 109). By linking the manager ID 109 to Employee_ID 109 we know Adam's manager is Lex De Hann.

EMPLOYEE_ID	FIRST_NAME	LAST_NAME	EMAIL	PHONE	HIRE_DATE	JOB_ID	SALARY	MANAGER_ID
100 Douglas	Grant	DGRANT	650.507.9844	23-JAN-08	SH_CLERK	2600	101
101 Adam	Fripp	AFRIPP	650.123.2234	10-APR-15	SH_MGR	8200	109
102 Jennifer	Whalen	JWHALEN	515.123.4444	17-SEP-13	AD_ASST	4400	108
103 Michael	Hartstein	MHARTSTE	515.123.5555	17-FEB-14	MK_MGR	13000	109
104 Pat	Fay	PFAY	603.123.6666	17-AUG-15	MK_REP	6000	103
105 Susan	Mavris	SMAVRIS	515.123.7777	07-JUN-12	HR_MGR	6500	109
106 Shelley	Higgins	SHIGGINS	515.123.8080	07-JUN-12	SA_MGR	12008	109
107 William	Gietz	WGIETZ	515.123.8181	07-JUN-12	SA_REP	8300	106
108 Steven	King	SKING	515.123.4567	17-JUN-13	AD_PRES	24000	108
109 Lex	De Haan	LDEHAAN	515.123.4569	13-JAN-11	AD_VP	17000	108
110 Bruce	Ernst	BERNST	590.423.4568	21-FEB-17	IT_MGR	6000	109

Figure 1.8 Self-referencing example

Summary

Chapter 1 covers the following:

- Introduction to the brief history of SQL and relational databases.
- Introduction to SQL standards.
- The basic terms of relational database management systems.
- Introduction to Oracle, SQL Server and MySQL versions.
- Displaying sample entity relationship diagram that used in this book.
- Defining one-to-one relationships.
- Defining one-to-many relationships.
- Defining many-to-many relationships.
- Defining self-referencing relationships.

Chapter 2

Data Types

You have learned in Chapter 1 that tables are consisted of many columns. When you design or modify databases it is very important to understand the different data types. There are three main data types: Characters, Numbers, and Date/Time.

Character Data Types

Table 2.1 Characters data types for the three database systems

Data Type	Oracle SQL	SQL SERVER	MySQL
Fixed-length Character	**CHAR**(n) Hold up to 2,000 characters	**CHAR**(n) Hold up to 8,000 characters	**CHAR**(n) Hold up to 255 characters
NCHAR for any language	**NCHAR**(n) Hold up to 2,000 characters	**NCHAR** Hold up to 4,000 characters	**NCHAR**(n) Hold up to 65,535 characters
variable-length character strings	**VARCHAR2**(n) Hold up to 4,000 characters	**VARCHAR**(n) Hold up to 8,000 characters **VARCHAR**(max) Hold up to 1,073 million characters	**VARCHAR**(n) Hold up to 255 characters
NVARCHAR2(n) for any language	**NVARCHAR2**(n) Hold up to 4,000 characters	**NVARCHAR** Hold up to 4,000 characters **NVARCHAR**(max) Hold up to 536 million characters	**NVARCHAR**(n) Hold up to 65,535 characters
			TINYTEXT Hold up to 255 characters
NTEXT for any language	**LONG** Variable width Hold up to 2 GB characters	**TEXT** **NTEXT** Hold up to 4,000 characters	**TEXT** Hold up to 65,535 characters
	RAW(n) Binary date Hold up to 2,000 bytes	**BINARY**(n) Fixed width binary date Hold up to 8,000 bytes	**MEDIUMTEXT** Hold up to 16 million characters
Character Large Object **NCLOB** for any language	**CLOB** **NCLOB** Hold up to 4G characters	**VARBINARY** Variable width binary date Hold up to 8,000 bytes	**LONGTEXT** 4G bytes
Binary Large Object	**BLOB** Hold up to 4G characters	**VARBINARY**(max) Variable width; Hold up to 2 GB	**LONGBLOB** Hold up to 4,294 million characters
		IMAGE Variable width; Hold up to 2 GB	**ENUM**(a, b, c, ...) List up to 65,535 values
			SET List up to 64 values

What is the difference between fixed-length characters and variable-length characters?

Fixed-length characters—When you create a fixed size field, like phone numbers, SSN, State, CHAR data type is a good choice.

Variable-length characters—Many fields have variable-length characters. When you create VARCHAR(30) or VARCHAR2(30) for first name field, for example, as first name length is different for each person you need to use VARCHAR or VARCHAR2 type. If a first name is "Peter", only 5 characters are stored in a table (5 bytes), not 30. If we use CHAR(30) for a first name field, than all the first names will be stored in 30 characters. Obviously, it will waste a lot of storage spaces.

Number Data Types

Table 2.2 Number data types for the three database systems

Data Type	Oracle SQL	SQL SERVER	MySQL
Small Integer	**NUMBER** (3) 0 to 255	**TINYINT** 0 to 255	**TINYINT** (n) −128 to 127 0 to 255 UNSIGNED
Median Integer	**NUMBER** (5)	**SMALLINT** −32,768 to 32767	**SMALLINT** (n) −32,768 to 32767 0 to 65,535 UNSIGNED
			MEDIUMINT (n) −8,388,608 to 8,388,608 0 to 16,772,215 UNSIGNED
Integer 32 bit	**NUMBER** (10)	**INT** −2,147,483,648 to −2,147,483,647	**INT** (n) −2,147,483,648 to −2,147,483,647 UNSIGNED
	NUMBER (38)	**BIGINT** −9,223,372,036,854,775,808 to 9,223,372,036,854,775,807	**BIGINT** (n) −9,223,372,036,854,775,808 to 9,223,372,036,854,775,807
		REAL Floating number −3.40E + 38 to 3.40E + 38	**FLOAT** (n, d) Small floating number n—maximum of digits d—decimal points
		SMALLMONEY −214,748.3648 to 214,748.3647 **MONEY** −922,337,203,685,477.5808 to 922,337,203,685,477.5807	**DOUBLE** (n, d) Large floating number n—maximum of digits d—decimal points
	NUMBER (p, s) **NUMERIC** (p, s) *p* from 1 – 38 *s* from −84 to 127	**DECIMAL** (p, s) **NUMERIC** (p, s) −10^38 + 1 to 10^38 – 1 *p* from 1 – 38 *s* from −84 to 127	**DECIMAL** (n, d) Stored as a string n—maximum of digits d—decimal points

NUMBER (p, s) (Oracle)

NUMERIC (p, s) (Oracle)

> p—precision
> s—scale

For example, NUMERIC (5, 2) including 3 digits before the decimal and 2 digits after the decimal.

DECIMAL (p, s) (T-SQL)

NUMERIC (p, s) (T-SQL)

> p—the maximum number of digits that can be stored (including all the digits from on the left and right of decimal point).
> s—the maximum number of digits that can be stored to the right of the decimal point.

Oracle Number Example

Datatype	Input Data	Stored Value
NUMBER	634,782.59	634782.59
NUMBER (8)	634,782.59	634783
NUMBER (8, 2)	634,782.59	634782.59
NUMBER (8, 1)	634,782.59	634782.5

Date and Time Data Types

Table 2.3 Date and time data types for the three database systems

Oracle SQL	SQL SERVER	MySQL
DATE Format: DD-MON-YY Example: 25-JAN-2017	**DATE** Format: YYYY-MM-DD Example: 2017-01-25	**DATE** Format: YYYY-MM-DD Example: 2017-01-25
TIMESTAMP (0) If we don't specify a precision then the timestamp defaults to six places.	**SMALLDATETIME** Format: YYYY-MM-DD HH:MI:SS **DATETIME** Format: YYYY-MM-DD HH:MI:SS [.mmm]	**DATETIME ()** Format: YYYY-MM-DD HH:MI:SS
	TIME HH:MI:SS.0000000	**TIME** (p) Format: HH:MI:SS
TIMESTAMP (3) DD-MM-YY HH:MI:SS	**TIMESTAMP** **Format:** YYYY-MM-DD HH:MI:SS	**TIMESTAMP** **Format:** YYYY-MM-DD HH:MI:SS
		YEAR ( ) Format: YY (70 to 69) 1970 to 2069 YYYY: 1901 to 2155

Boolean Data Type

Table 2.4 Boolean data types for the three database systems

Data Type	Oracle SQL	SQL SERVER	MySQL
Boolean	**CHAR**(1) (0 or 1)	**BIT** 0, 1 and NULL	**BOOLEAN, BOOL** 0 or 1; Not NULL

Summary

Chapter 2 covers several data types for the three database systems.

- Character data types
- Number data types
- Date and time data types in the three database systems
- Boolean data type

Although this chapter is short but it takes time to get familiar with all those date types. When you study Chapter 5 "Creating Databases and Tables" you will use different data types for columns.

In the next chapter we will install Oracle 12c, SQL Server 2016 and MySQL 5.7 database systems.

Chapter 3

Installation of Oracle, SQL Server and MySQL

Before we run SQL commands we need to install relational database management systems. This chapter covers how to install Oracle 12c, SQL Server 2016 and MySQL 5.7.

Minimum System Requirements

Table 3.1 System requirements

Oracle 12C	SQL Server 2016	MySQL 5.7
Hard Disk: 10 GB	**Hard Disk**: 6 GB A DVD drive is required for installation from disc.	**Hard Disk**: 8 GB
RAM: 2 GB 1 GB of space in the tmp directory.	.NET Framework 4.6	**RAM**: 2 GB
	Recommended RAM	**Operating System**
Operating System 32-bit: Windows 8 (Pro and Enterprise editions) Windows 7 (Professional, Enterprise, Ultimate editions) Windows Server 2008	Express Editions: 1 GB All other editions: At least 4 GB **Processor**: x64 Processor **Operating System** SQL Server Enterprise 　　Windows Server 2016 　　Windows Server 2012	Windows 32-bit and 64-bit Linux Mac OS X
64-bit: Windows 8 (Pro and Enterprise editions) Windows 7 (Professional, Enterprise, Ultimate editions) Windows Server 2012 Windows Server 2008 R2 Windows Server 2008	SQL Server Standard 　　Windows Server 2016 　　Windows Server 2012 　　Windows 10 　　Windows 8.1	
Linux	SQL Server Web and Express: 　　Windows Server 2016 　　Windows Server 2012	
	SQL Server Developer: 　　Windows Server 2016 　　Windows Server 2012 　　Windows 10 　　Windows 8.1 　　Windows 8	

Installation of Oracle 12c

- Download Oracle Database 12*c* Release 2 from the Oracle Web site:

  http://www.oracle.com/technetwork/database/enterprise-edition/downloads/database12c-win64-download-2297732.html

Figure 3.1 Oracle 12c downloads

- After downloading and decompressing Windows x64 files, make sure that two folders are at the same location:

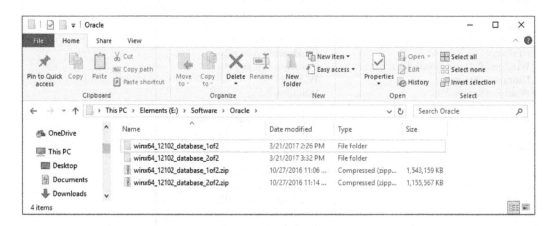

Figure 3.2 Downloaded files

- There are extra steps for Windows 7 PCs:
  1. Open the winx64_12c_database_2of2 directory
  2. Copy all the files under\winx64_12c_database_2of2\database\stage\Components directory
  3. Paste all the files to\winx64_12c_database_1of2\database\stage\Components

- Go to\winx64_12102_database_1of 2 and run the **setup.exe** file:

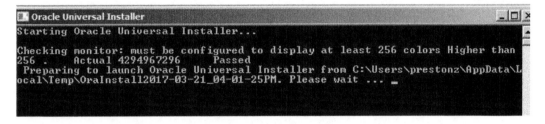

Figure 3.3 Running setup file

- Choose the languages:

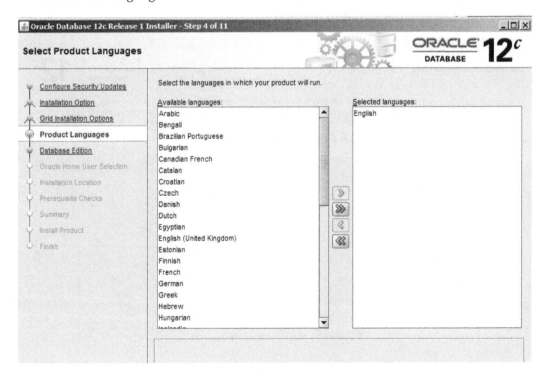

Figure 3.4 Choosing a language

- Select the database edition:

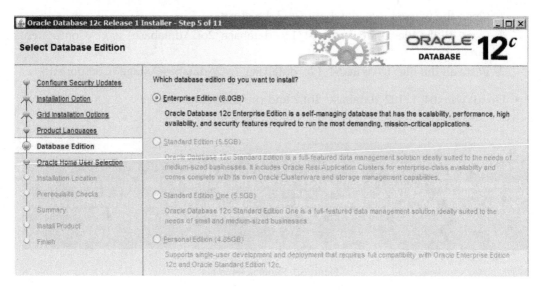

Figure 3.5 Selecting a database edition

- Choose **"Use Windows Built-in Account"**:

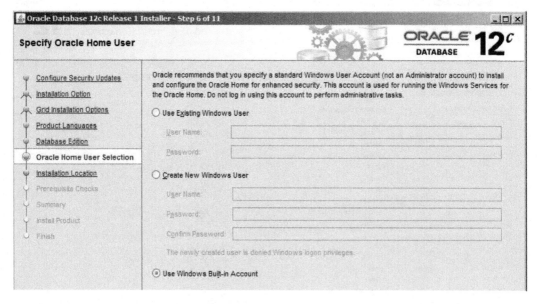

Figure 3.6 Windows built-in account

- Click "**Yes**" for the following warning message:

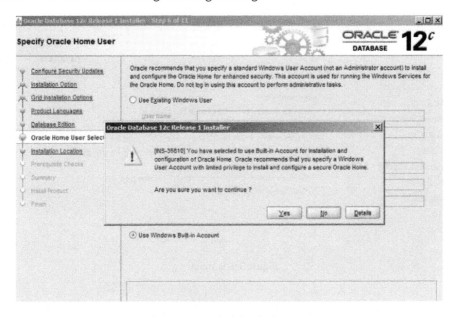

Figure 3.7 Warning message

- Choose Installation Location:

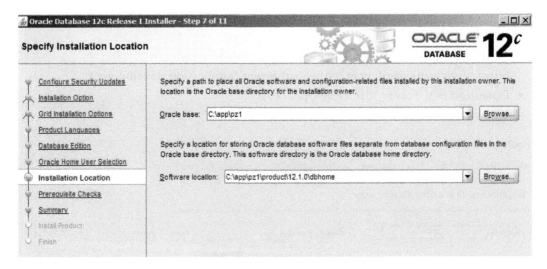

Figure 3.8 Installation location

- After summary page click "**Next**" to install Oracle 12c:

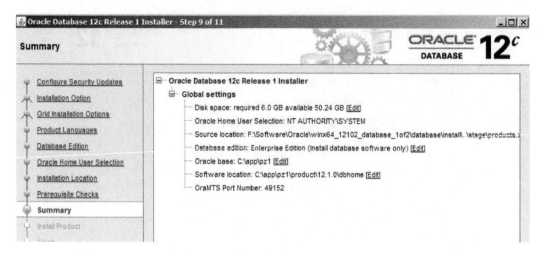

Figure 3.9 Summary

- After the installation you can install Oracle SQL Developer. The step by step instructions is in the next chapter.

Installation of SQL Server 2016

The SQL Server 2016 Installation is straightforward. Every installation creates one SQL Server instance on your computer.

- Go to SQL Server 2016 Developer Edition download page:

  https://www.microsoft.com/en-us/sql-server/application-development

- Download SQL Server 2016 Developer (x64).

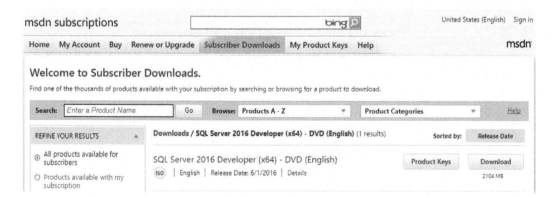

Figure 3.10 SQL Server 2016 downloads

- Click **setup.exe** to run the installation file.

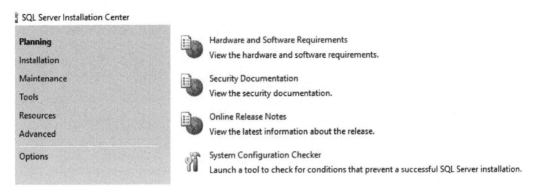

Figure 3.11 Running setup file

- Select **Developer** edition.

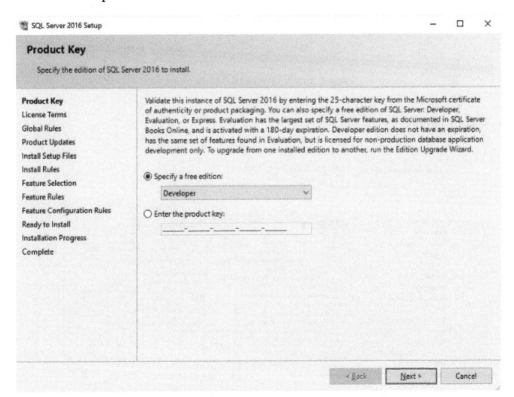

Figure 3.12 Selecting developer edition

- Accept the license terms.

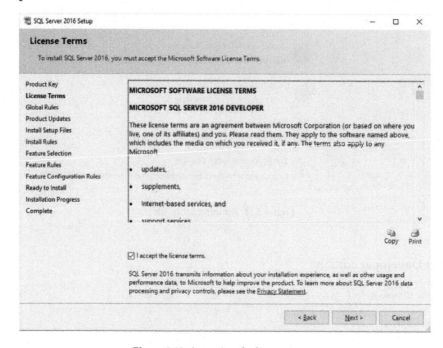

Figure 3.13 Accepting the license terms

- Select Features:

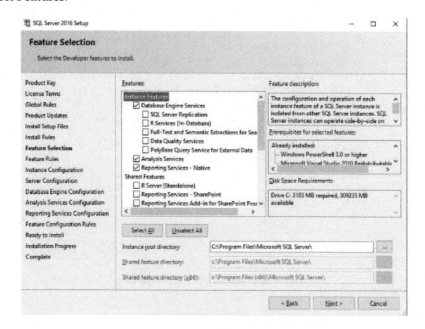

Figure 3.14 Selecting features

- SQL Server issues a default instance name: SQL 2016. Any previously installed instances will be displayed here.

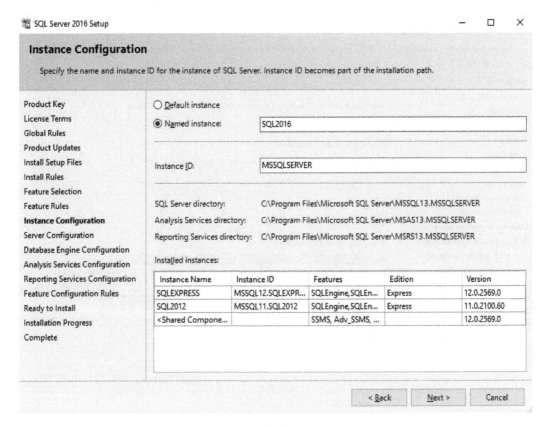

Figure 3.15 Default instance name

- Click **"Add Current User"** to set up an administrator:

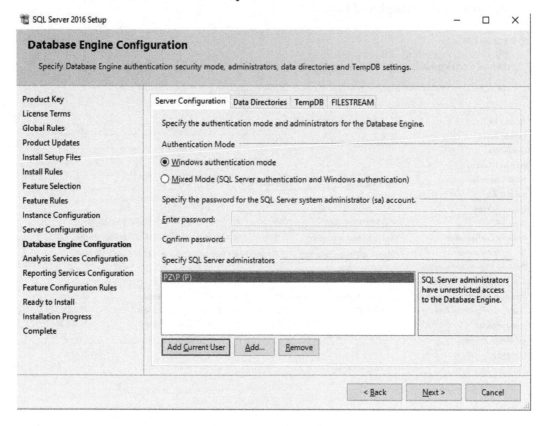

Figure 3.16 Setting up an administrator

- Click **Next** button then click **Install** button.

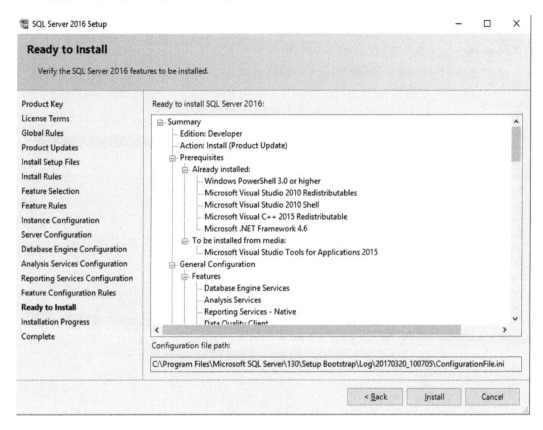

Figure 3.17 Summary

- Installation is completed.

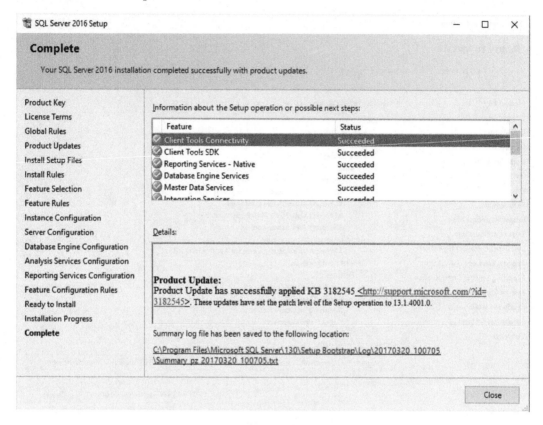

Figure 3.18 Installation is done.

- After the installation you can install SQL Server Management Studio. The step by step instructions is in the next chapter.

Installation of MySQL

- Go to MySQL installer page: https://dev.mysql.com/downloads/installer/

Note: MySQL Installer is 32 bit, but will install both 32 bit and 64 bit binaries.

Online Documentation

· MySQL Installer Documentation and Change History

Please report any bugs or inconsistencies you observe to our Bugs Database.
Thank you for your support!

Generally Available (GA) Releases	Development Releases

MySQL Installer 5.7.17

Select Operating System: Looking for previous GA versions?

Microsoft Windows	▼

Windows (x86, 32-bit), MSI Installer	5.7.17	1.7M	Download
(mysql-installer-web-community-5.7.17.0.msi)		MD5: d#80081cd386da03248c4#b4bae37758 | Signature	
Windows (x86, 32-bit), MSI Installer	5.7.17	386.6M	Download

Figure 3.19 MySQL downloads

- Choose **MySQL Enterprise Edition** or **Standard Edition**.

Oracle Software Delivery Cloud

Need Help? Contact Software Delivery Customer Service

To add items to your Download Queue, enter the Oracle Product or Release into the type-ahead field below, then select from the list of available platforms. ∧
The title will be displayed in the Download Queue. Repeat this step for all titles you wish to download. Once complete, click 'Continue'.

Filter Products by ☑ Programs   ☑ Linux/OVM/VMs   ☐ Self-Study Courseware   ☐ 1-Click Offerings

Search by All ∨ MySQL Standard Edition   Select Platform ∨

Download Qu			Continue
	☐ Apple Mac OS X (Intel) (64-bit)		
Selected Item	☐ FreeBSD - x86	**Platform**	
	☐ Linux x86		
Product: MySQL Enterpri	☐ Linux x86-64	Microsoft Windows x64 (64-bit)	
	☐ Microsoft Windows (32-bit)		
	☑ Microsoft Windows x64 (64-bit)		
	☐ Oracle Solaris on SPARC (64-bit)		
	☐ Oracle Solaris on x86-64 (64-bit)		

Select Cancel

Remove All	Continue

Figure 3.20 Choosing edition to download

- Select **MySQL Standard Edition for Microsoft Windows x64 (64-bit).**

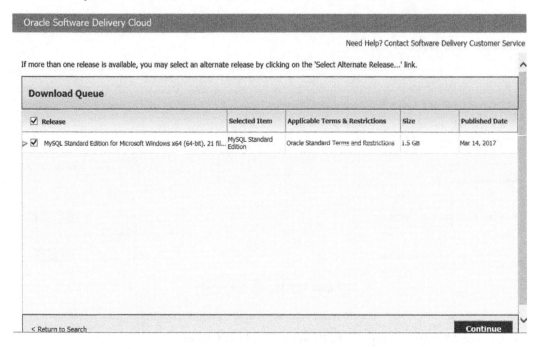

Figure 3.21 Selected program

- Accept the license terms.

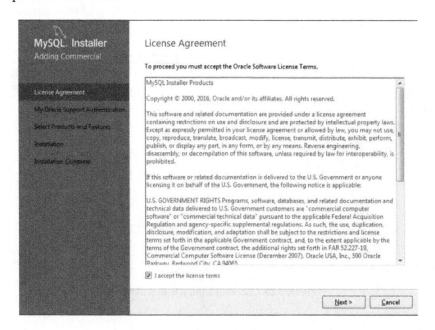

Figure 3.22 Accepting the license terms

- If you do not use Oracle Support select '**No**'.

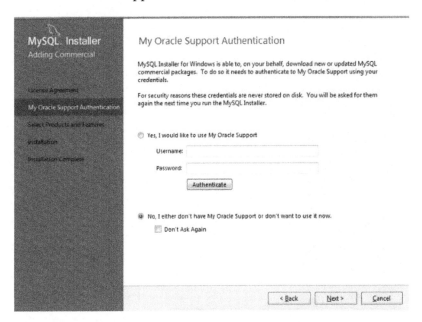

Figure 3.23 Choosing the support option

- You can select MySQL documentation and samples:

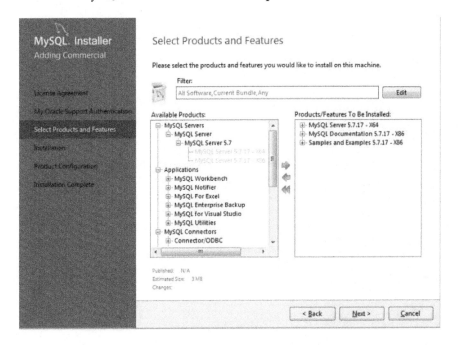

Figure 3.24 Selecting documents or samples

- Installation is ready to go.

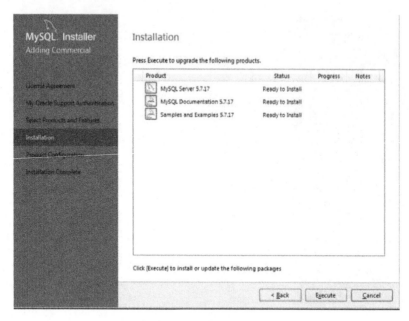

Figure 3.25 Installation is ready

- Keep default Server Configuration Setting:

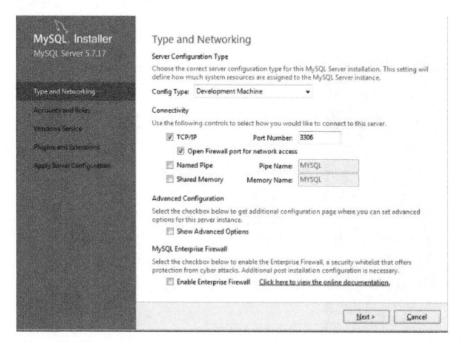

Figure 3.26 Default server configuration

- Enter root account password. Please remember this password as you will use it to login to MySQL server. If you want to add users you can click **"Add User"** button.

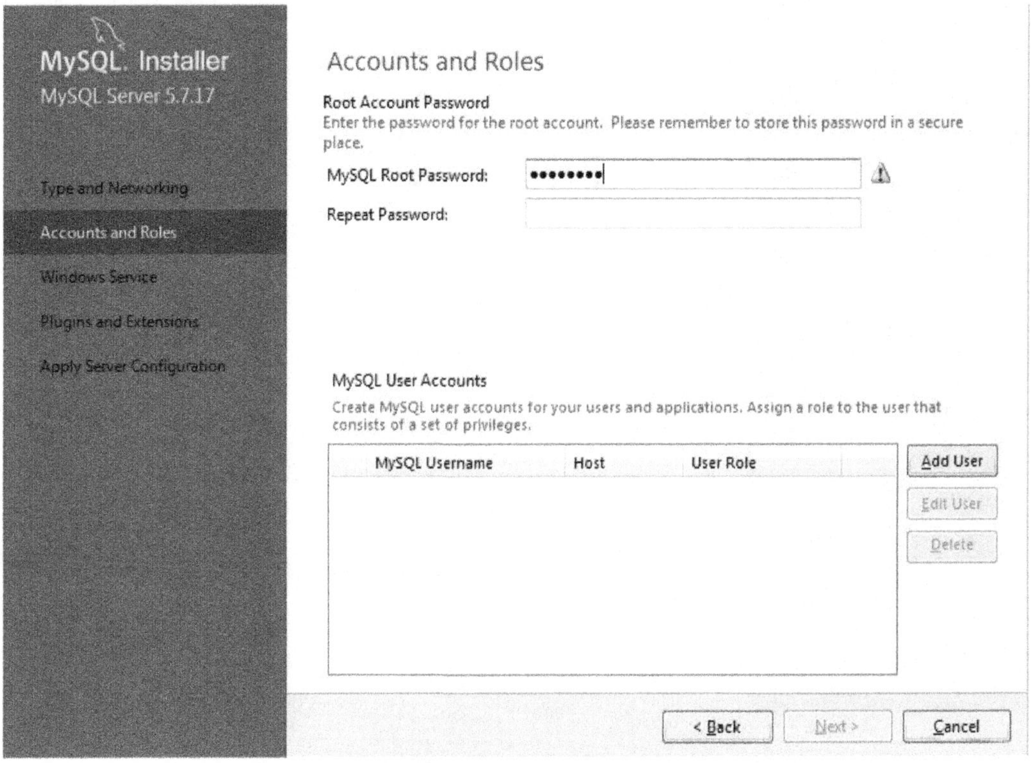

Figure 3.27 Entering account password

- Enter the password and click **"Check"** button to see if it is working.

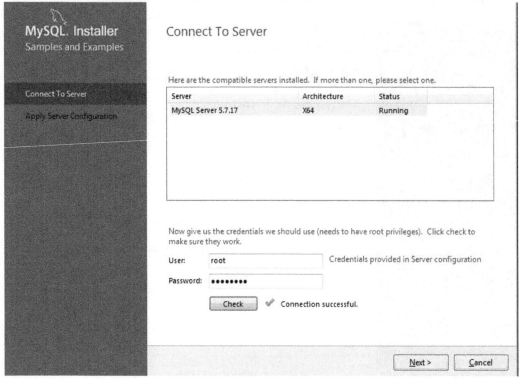

Figure 3.28 Testing the password

- After the installation you can install MySQL Workbench. The step by step instructions is in the next chapter.

Summary

Chapter 3 covers the following:

- Minimum System Requirements
- How to install Oracle 12c
- How to install SQL Server 2016
- How to install MySQL 5.7

In the next chapter you are going to install development tools for the three database systems.

Exercise

3.1

Install Oracle 12c or SQL Server 2016 or MySQL 5.7 on your computer following the instructions in this chapter. If you want to test SQL statements for the three database systems then install them all on your computer.

Chapter 4

Database Development Tools

There are many database development tools available: Command line tools and graphic user interface tools. Command Line Tools including Oracle SQL Plus and MySQL Command Line Client. Graphic User Interface Tools include Oracle SQL Developer, SQL Server Management Studio and MySQL Workbench.

Command Line Tools

Oracle SQL Plus

- Go to Start -> Oracle-OraDB12Home1 -> SQL Plus

Figure 4.1 Starting SQL Plus

- To access a built-in database HR enter the username and password of the HR schema.

    Enter user-name: *hr*

    Enter password: *xx* (you can reset the password if you forgot it)

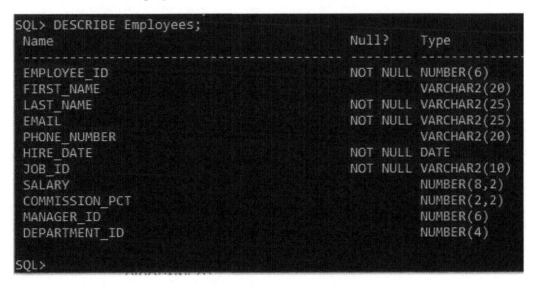

```
SQL Plus

SQL*Plus: Release 12.1.0.2.0 Production on Mon Apr 10 10:38:21 2017

Copyright (c) 1982, 2014, Oracle.  All rights reserved.

Enter user-name: hr
Enter password:
Last Successful login time: Mon Apr 10 2017 10:32:50 -04:00

Connected to:
Oracle Database 12c Enterprise Edition Release 12.1.0.2.0 - 64bit Production
With the Partitioning, OLAP, Advanced Analytics and Real Application Testing options

SQL>
```

Figure 4.2 SQL Prompt

- To see the structure of the Employees table enter:

    DESCRIBE Employees;

```
SQL> DESCRIBE Employees;
 Name                                      Null?    Type
 ----------------------------------------- -------- ----------------
 EMPLOYEE_ID                               NOT NULL NUMBER(6)
 FIRST_NAME                                         VARCHAR2(20)
 LAST_NAME                                 NOT NULL VARCHAR2(25)
 EMAIL                                     NOT NULL VARCHAR2(25)
 PHONE_NUMBER                                       VARCHAR2(20)
 HIRE_DATE                                 NOT NULL DATE
 JOB_ID                                    NOT NULL VARCHAR2(10)
 SALARY                                             NUMBER(8,2)
 COMMISSION_PCT                                     NUMBER(2,2)
 MANAGER_ID                                         NUMBER(6)
 DEPARTMENT_ID                                      NUMBER(4)

SQL>
```

Figure 4.3 Testing a SQL statement

- Enter "**EXIT**" to leave the SQL prompt.

MySQL Command Line Client

- Go to Start -> MySQL -> MySQL 5.7 Command Line Client
  (The second MySQL 5.7 Command Line Client is for Unicode)

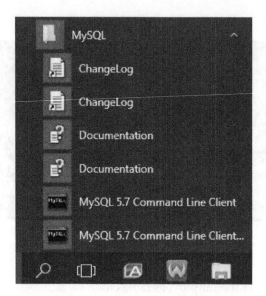

Figure 4.4 Starting MySQL Command Line Client

- Enter MySQL root password (You setup a password when you install the MySQL 5.7)

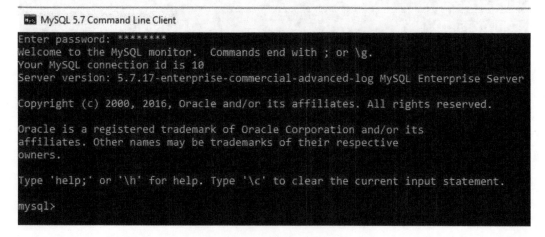

Figure 4.5 MySQL prompt

- Another way to start MySQL prompt:

     type C:\Program Files\MySQL\MySQL Server 5.7\bin\mysql –u root –p

```
C:\Program Files\MySQL\MySQL Server 5.7\bin>mysql -u root -p
Enter password: ********
Welcome to the MySQL monitor.  Commands end with ; or \g.
Your MySQL connection id is 9
Server version: 5.7.17-enterprise-commercial-advanced-log MySQL Enterprise Server

Copyright (c) 2000, 2016, Oracle and/or its affiliates. All rights reserved.

Oracle is a registered trademark of Oracle Corporation and/or its
affiliates. Other names may be trademarks of their respective
owners.

Type 'help;' or '\h' for help. Type '\c' to clear the current input statement.

mysql>
```

Figure 4.6 Displaying MySQL prompt in other way

- Enter "**SHOW databases**" at the MySQL prompt.

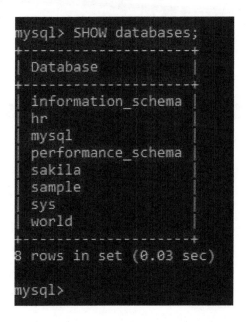

```
mysql> SHOW databases;
+--------------------+
| Database           |
+--------------------+
| information_schema |
| hr                 |
| mysql              |
| performance_schema |
| sakila             |
| sample             |
| sys                |
| world              |
+--------------------+
8 rows in set (0.03 sec)

mysql>
```

Figure 4.7 Testing a SQL statement

Graphic User Interface Tools

Installation of Oracle SQL Developer

- Download Oracle SQL Developer at the following link:
  http://www.oracle.com/technetwork/developer-tools/sql-developer/downloads/
  index.html
- Starts Oracle SQL Developer:

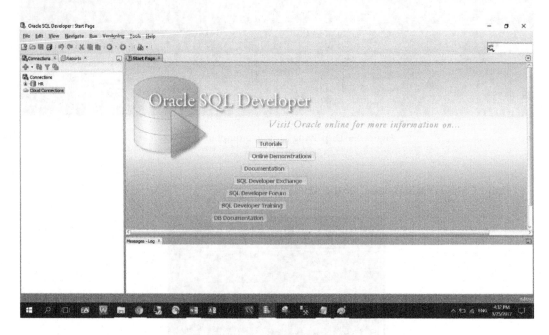

Figure 4.8 Starting Oracle SQL Developer

- Oracle has many build-in schemas. Here we use Oracle build-in HR schema to make a connection:

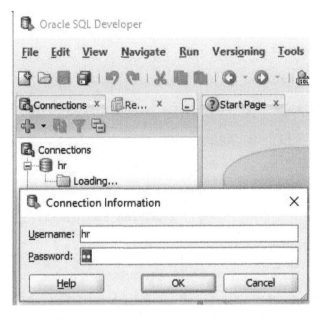

Figure 4.9 Entering username and password

- Enter password then click **OK** button.
- SQL Developer opens Connections pane on the left and SQL worksheet on the right.

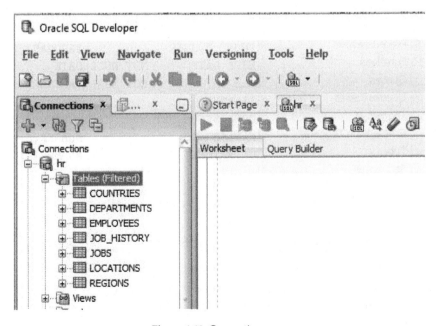

Figure 4.10 Connection pane

- Enter "**DESCRIBE Employees;**" on the Worksheet then click the **Run Statements** button (the green triangle). You can see that we get the same result as the SQL Plus command line.

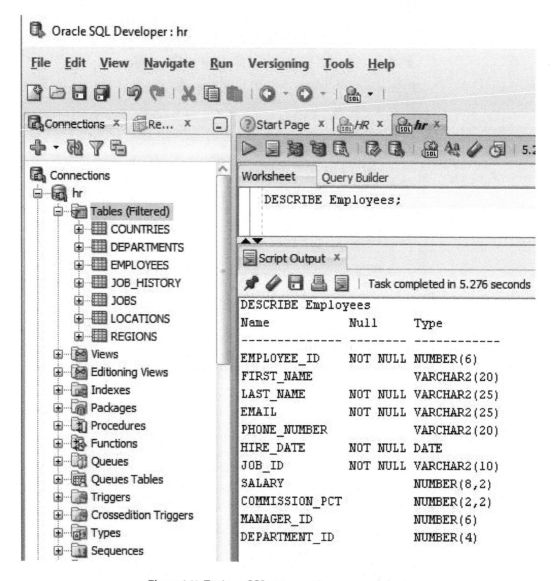

Figure 4.11 Testing a SQL statement in query worksheet

Installation of SQL Server Management Studio

- From SQL Server 2016 the server and the management studio are installed separated.
- Go to SQL Server 2016 Management Studio download page:

  https://docs.microsoft.com/en-us/sql/ssms/download-sql-server-management-studio-ssms

Figure 4.12 SQL Server Management Studio downloads

- After the download start the **setup.exe** file.

Figure 4.13 Starting setup file

- When the installation is done let us start SQL Server Management Studio:

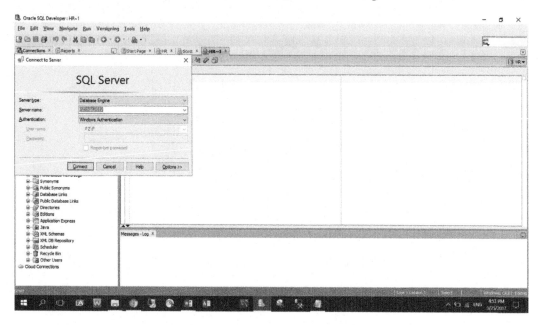

Figure 4.14 Starting SQL Server Management Studio

- Check the Server Name then click **Connect** button. The SQL Server Management Studio has Object Explorer pane on the left. You can navigate through databases, tables, columns, or other types of objects. Click **New Query** button to enter query statements.

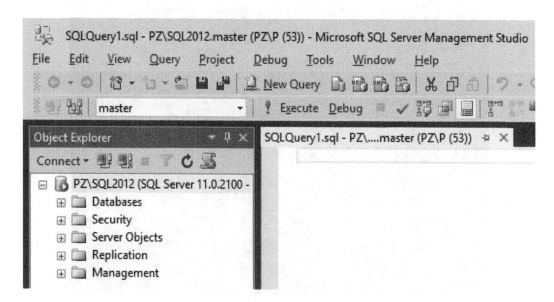

Figure 4.15 Master database

- The "**Master**" database will be selected by default. Make sure to select a database that you are working on. Or you can type "**Use database_name;**" command on the SQL worksheet before a query statement. To run a query click "**! Execute**" button.

Installation of MySQL Workbench

- Download MySQL Workbench at:

  https://dev.mysql.com/downloads/workbench/

Figure 4.16 MySQL Workbench downloads

- After the installation MySQL Workbench starts:

Figure 4.17 Starting MySQL Workbench

- Click "**Local instance MySQL 57**" and enter the password (You setup the password when you install MySQL). You are now ready to use MySQL Workbench.

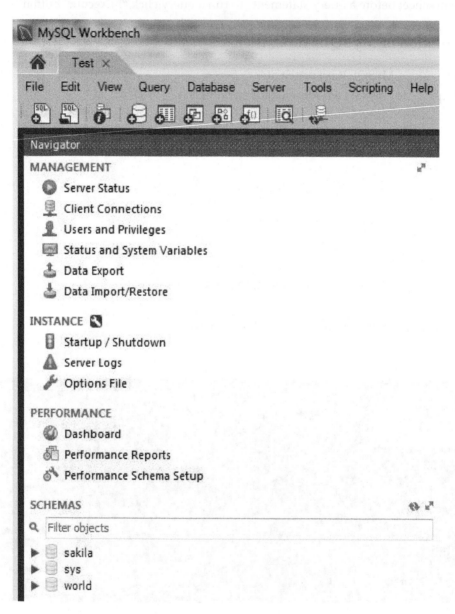

Figure 4.18 Navigation pane

- Click the arrow icon on the right side of SCHEMAS to move the SCHEMAS to the top of Navigator pane:

Figure 4.19 Displaying Schemas

- To open a SQL worksheet click **"Create a new SQL tab for executing queries"** above the Navigator pane.
- Enter **"SHOW databases;"** command and click the Literning icon to run the query.

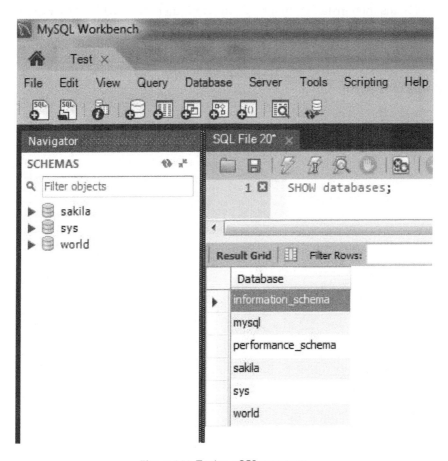

Figure 4.20 Testing a SQL statement

Summary

Chapter 4 covers the following:

- Introduction to SQL Plus
- Introduction to MySQL Command Line Client
- Installation of Oracle SQL Developer
- Installation of SQL Server Management Studio
- Installation of MySQL Workbench

Now you are ready to create databases and tables in the next chapter.

Exercises

4.1

Install Oracle SQL Developer, SQL Server Management Studio and MySQL Workbench on your computer based on your need. If you want to test SQL statements in the three database systems then install them all on your computer. Test each tool by running a SQL command.

4.2

View the databases and tables in Oracle SQL Developer, SQL Server Management Studio and MySQL Workbench.

Data Definition Language (DDL)

We have installed the database systems and development tools in Chapter 3 and Chapter 4. Now we are ready to create databases and tables.

SQL statements are divided into three main groups:

- Data Definition Language (DDL)
- Data Manipulation Language (DML)
- Data Control Language (DCL)

Below is the statement summary for DDL, DML and DCL.

Table 5.1 DDL, DML and DCL

Language	Statements	
Data Definition Language (DDL)	**CREATE**	– To Create objects in the database
	DROP	– To delete objects from the database
	ALTER	– To change database structure
	RENAME	– To rename a database object
	TRUNCATE	– To remove all records from a table
Data Manipulation Language (DML)	**SELECT**	– To retrieve data from a database
	INSERT INTO	– To insert data into a table
	UPDATE... SET	– To update data in a table
	DELETE FROM	– To deletes all rows from a table
Data Control Language (DCL)	**GRANT**	– To grant privileges to a user
	REVOKE	– To revoke privileges from a user

We cover Data Definition Language (DDL) in this chapter. Data Manipulation Language (DML) and Data Control Language (DCL) will be discussed in the next chapter.

Data Definition Language Statements

Creating a Database

We can create a database in two ways:

1. Using SQL Command
2. Using Graphic User Interface (GUI) Tools

1. Using SQL Commands to Create a Database

Syntax

> **CREATE DATABASE** Database_Name;

Steps to Create a Database in Oracle:

1. Login to Oracle SQL Plus as system user
2. At the sql prompt enter:

> sql> **CREATE USER** TEST_DB IDENTIFIED BY pw;

Note: An Oracle's user name acts as database name. We created a user "TEST_DB" with password "pw".

3. Grant privileges to the user.

> Sql> **GRANT** CONNECT, DBA TO TEST_DB;

4. Create all the objects like tables under the user.

Steps to Create a Database in SQL Server:

1. In the SQL Management Studio query worksheet enter:

> **CREATE DATABASE** TEST_DB;

2. Run the query and refresh *Connect* pane.
3. The TEST_DB database is created in the *Connect* pane.

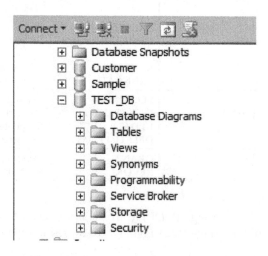

Figure 5.1 Creating a database in SQL Server

Steps to Create a Database in MySQL:

1. In the MySQL Workbench query worksheet enter:

> **CREATE DATABASE** TEST_DB;

2. Run the query and refresh *SCHEMAS*.
3. The test_db database is created in the *Navigator* pane.

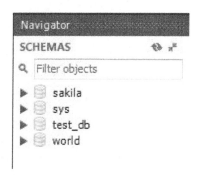

Figure 5.2 Creating a database in MySQL

2. Using GUI Tools to Create a Database

Creating a Database in GUI (Oracle)

- Using **Database Configuration Assistant** to create a database after installation.

Figure 5.3 Database Configuration Assistant

Creating a Database in GUI (SQL Server)

- Right click **Database** and enter **Database Name:** Car
- Click OK

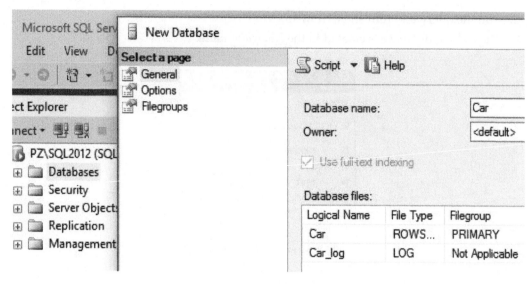

Figure 5.4 Creating a database in GUI (SQL Server)

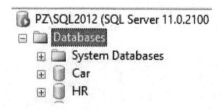

Figure 5.5 The database "CAR" is created

Creating a Database in GUI (MySQL)

- Right click any schema and choose *Create Schema*

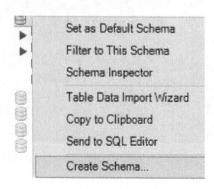

Figure 5.6 Creating a database in GUI (MySQL)

- Enter schema name 'hr' and click **Apply**

Figure 5.7 Entering a database (schema) name

Dropping a Database

Description: Deletes a Database. Be careful to drop database or drop table. If you do not backup your database you may lose important data!

Syntax (Oracle)

> **DROP USER** User_Name;

Syntax (T-SQL & MySQL)

> **DROP DATABASE** Database_Name;

Table 5.2 Dropping database

Oracle	T-SQL	MySQL
DROP USER TEST_DB;	**DROP DATABASE HR;** Command(s) completed successfully.	**DROP DATABASE hr;** 0 row(s) affected.

Creating a Table

We can create tables in two ways too:

1. Using SQL Command
2. Using Graphic User Interface Tools

Note:

- All the Oracle tables in this book are created in a user account.
- All the T-SQL tables are created in HR database.
- All the MySQL tables are created in hr schema (database).

1. Using SQL Commands to Create a Table

Syntax

> **CREATE TABLE** Table_Name;

Creating Regions Table

Table 5.3 Creating Regions table commands

Oracle SQL	T-SQL & MySQL

CREATE TABLE Regions
(
   Region_id   NUMBER NOT NULL,
   Region_name VARCHAR2(25),
   PRIMARY KEY (Regions_ID)
);
	CREATE TABLE regions
	(
	   region_id NUMBER NOT NULL,
	   region_name VARCHAR(25),
	   PRIMARY KEY (Regions_ID)
	);

Creating Country Table

Table 5.4 Creating Country table commands

Oracle SQL	T-SQL & MySQL

CREATE TABLE Country
(
   Country_id   CHAR (2) NOT NULL,
   Country_name VARCHAR2(40),
   Region_id   NUMBER,
   PRIMARY KEY (country_ID),
   CONSTRAINT FK_RegCountry
   FOREIGN KEY (region_id)
   REFERENCES Regions (Regins_ID)
);
	CREATE TABLE Country
	(
	   country_id CHAR (2) NOT NULL,
	   country_name VARCHAR(40),
	   region_id smallint,
	   PRIMARY KEY (country_ID),
	   CONSTRAINT FK_RegCountry
	   FOREIGN KEY (region_id)
	   REFERENCES Regions (Regins_ID)
	);

Creating Departments Table

Table 5.5 Creating Departments table commands

Oracle SQL	T-SQL & MySQL

CREATE TABLE Departments
(
   Dept_id   NUMBER (4) NOT NULL,
   Dept_name VARCHAR2 (30),
   Manager_id NUMBER (6),
   Location_id NUMBER (4),
   PRIMARY KEY (dept_ID),
   CONSTRAINT FK_LocDept
   FOREIGN KEY (location_id)
   REFERENCES Locations (location_ID)
);
	CREATE TABLE Departments
	(
	   dept_id NUMBER (4) NOT NULL,
	   dept_name VARCHAR (30),
	   manager_id NUMBER (6),
	   location_id NUMBER (4),
	   PRIMARY KEY (dept_ID),
	   CONSTRAINT FK_LocDept
	   FOREIGN KEY (location_id)
	   REFERENCES Locations (location_ID)
	);

Creating Employees Table

Table 5.6 Creating Employees table commands

Oracle SQL	T-SQL & MySQL

CREATE TABLE Employees
(
   Employee_ID   NUMBER(6) NOT NULL,
   First_NAME   VARCHAR2(20) ,
   Last_Name   VARCHAR2(25),
   Email   VARCHAR2(25),
   Phone   VARCHAR2(20),
   Hire_Date   DATE,
   Job_ID   VARCHAR2(10),
   Salary   NUMBER(6, 2),
   Manager_ID   NUMBER(6),
   Dept_ID   NUMBER(4),
   PRIMARY KEY (Employee_ID),
   CONSTRAINT FK_DepEmp
   FOREIGN KEY (Dept_ID)
   REFERENCES Dept (Dept_ID)
);
	CREATE TABLE employees
	(
	   Employee_ID int NOT NULL,
	   First_NAME varchar(20),
	   Last_Name varchar(25),
	   Email varchar(25),
	   Phone varchar(20),
	   Hire_Date date,
	   Job_ID varchar(10),
	   Salary decimal(6, 2),
	   Manager_ID decimal(6,0),
	   Dept_ID smallint,
	   PRIMARY KEY (Employee_ID),
	   CONSTRAINT FK_DepEmp
	   FOREIGN KEY (Dept_ID)
	   REFERENCES Dept (Dept_ID)
	);

Creating Locations Table

Table 5.7 Creating Locations table commands

Oracle SQL	T-SQL & MySQL

CREATE TABLE Locations
(
   Location_id   NUMBER (4) NOT NULL,
   Street_address   VARCHAR2(40),
   Postal_code   VARCHAR2(12),
   City   VARCHAR2(30),
   State_province   VARCHAR2(25),
   Country_id   CHAR(2),
   PRIMARY KEY (Location_ID),
   CONSTRAINT FK_CountryLoc
   FOREIGN KEY (country_id)
   REFERENCES country (country_ID)
);
	CREATE TABLE Locations
	(
	   location_id NUMBER (4) NOT NULL,
	   street_address VARCHAR(40),
	   postal_code VARCHAR(12),
	   city VARCHAR(30),
	   state_province VARCHAR(25),
	   country_id CHAR(2),
	   PRIMARY KEY (Location_ID),
	   CONSTRAINT FK_CountryLoc
	   FOREIGN KEY (country_id)
	   REFERENCES country (country_ID)
	);

Creating Job Table

Table 5.8 Creating Job table commands

Oracle SQL	T-SQL & MySQL

CREATE TABLE Job
(
   Job_id   VARCHAR2 (10) NOT NULL,
   Job_title   VARCHAR2 (35),
   Min_salary   NUMBER (6),
   Max_salary   NUMBER (6),
   PRIMARY KEY (job_ID)
);
	CREATE TABLE Job
	(
	   job_id VARCHAR (10) NOT NULL,
	   job_title VARCHAR (35),
	   min_salary NUMBER (6),
	   max_salary NUMBER (6),
	   PRIMARY KEY (job_ID)
	);

2. Using GUI Tools to Create a Table

1) Oracle

* Right click the *Table* and select *New Table.*

Figure 5.8 Creating a new table in Oracle

* Enter table name "Customer" and column names. Choose data types and sizes.

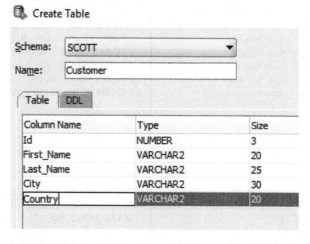

Figure 5.9 Entering the table name and the column names

2) SQL Server

- Right click the *Table* and select *New Table.*

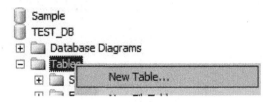

Figure 5.10 Creating a table in SQL Server

- Enter column names. Choose data types and sizes.

	Column Name	Data Type
	ID	int
	FirstName	varchar(15)
	LastName	varchar(15)
	Title	varchar(20)
	HireDate	date
	Salary	decimal(7, 2)
	DeptNo	smallint

Figure 5.11 Entering the column names

3) MySQL

- Right click the *Table* and select *Create Table.*

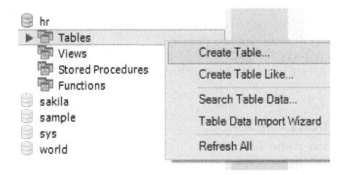

Figure 5.12 Creating a table in MySQL

- Enter table name and column names. Choose data types and sizes.

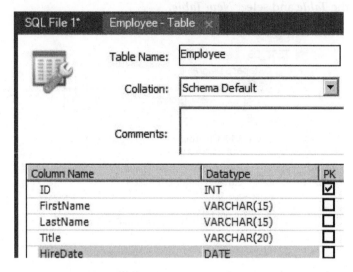

Figure 5.13 Entering the table name and the column names

Using Data from an Existing Table to Create a Table

Syntax (Oracle & MySQL)

> **CREATE TABLE** Table_Name AS
> SELECT ... FROM

Syntax (T-SQL)

> **SELECT ... INTO** Table_Name
> **FROM** Original_Table

Question 1: Write a query to create a table Emp using the Employees table and having the employees who were hired after January 1st 2017.

Answer in Oracle SQL:

> **CREATE TABLE** Emp **AS**
> SELECT First_name, Last_Name
> FROM Employees
> WHERE Hire_date >= '01-JAN-2017';

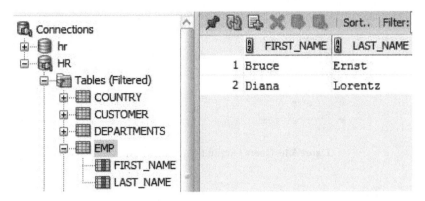

Figure 5.14 Query Output for question 1 (Oracle)

Answer in T-SQL:

> **SELECT** First_Name, Last_Name **INTO** Emp
> FROM Employees
> WHERE Hire_date >= '2017-01-01';

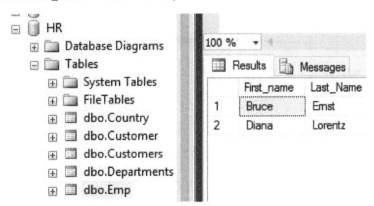

Figure 5.15 Query Output for question 1 (SQL Server)

Answer in MySQL:

> **CREATE TABLE** Emp As
> SELECT First_name, Last_Name
> FROM Employees
> WHERE Hire_date >= '2017-01-01';

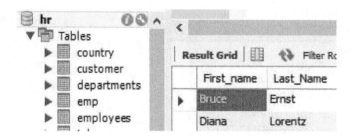

Figure 5.16 Query Output for question 1 (MySQL)

Renaming a Table

Syntax (Oracle)

> **RENAME** old_table_name **TO** new_table_name;

Syntax (MySQL)

> **RENAME TABLE** old_table_name **TO** new_table_name;

Question 2: Rename Departments table to Dept.

Answer (Oracle):

> **RENAME** Departments **TO** Dept;

Answer (MySQL):

> **RENAME TABLE** Departments **TO** Dept;

Oracle	MySQL
Tables (Filtered)	Tables
COUNTRY	country
CUSTOMER	customer
DEPT	dept
EMP	employees
EMPLOYEES	job
JOB	locations
LOCATIONS	orders
ORDERS	regions
REGIONS	state

Figure 5.17 Two tables are renamed

Renaming a Table in GUI Tools

1) Oracle

- Right click *Departments* table and select ***Table->Rename***

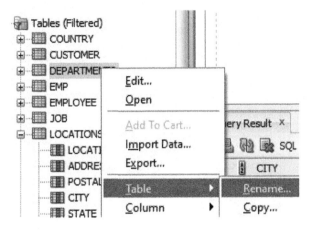

Figure 5.18 GUI renaming name in Oracle

2) SQL Server

- Right click the *dbo.Departments* and select ***Rename.***

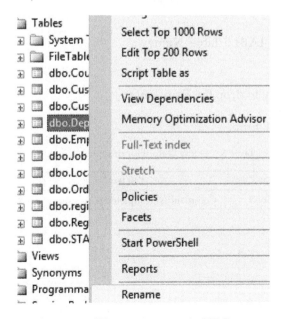

Figure 5.19 GUI renaming name in SQL Server

3) MySQL

- Right click *departments* table and select *Alter Table...*

Figure 5.20 GUI renaming name in MySQL

Dropping a Table

Description: Deletes a table from a database.

Syntax

> **DROP TABLE** Table_Name;

Question: Write a query to drop the Emp Table.

Answer: **DROP TABLE** Emp;

Table 5.9 Query Output for dropping command

Oracle SQL	T-SQL	MySQL
table EMP dropped.	Command(s) completed successfully.	0 row(s) affected.

Truncating a Table

Description: To remove all records from a table. The operation can't be rolled back.

Syntax

TRUNCATE TABLE Table_Name;

Question: Write a query to remove all the records in Customers table.

Answer: **TRUNCATE TABLE** Customers;

Table 5.10 Query Output for TRUNCATE command

Oracle SQL	T-SQL	MySQL
table CUSTOMERS truncated.	Command(s) completed successfully.	0 row(s) affected.

Altering a Table (Modifying a Column)

Syntax

ALTER TABLE MODIFY Column Type (Oracle & MySQL)

ALTER TABLE ALTER COLUMN Column Type (T-SQL)

Question: Write a query to Modify Column Salary from Number (6, 2) to Number (7, 2).

Answer (Oracle):

ALTER TABLE EMPLOYEES **MODIFY** SALARY NUMBER (7, 2);

	COLUMN_NAME	DATA_TYPE
1	ID	NUMBER(6,0)
2	FIRSTNAME	VARCHAR2(15 BYTE)
3	LASTNAME	VARCHAR2(15 BYTE)
4	TITLE	VARCHAR2(20 BYTE)
5	HIREDATE	DATE
6	SALARY	NUMBER(7,2)
7	DEPTNO	NUMBER(3,0)
8	COMMISION	NUMBER(6,2)

Figure 5.21 Altering a table in Oracle

Answer (T-SQL):

ALTER TABLE EMPLOYEES **ALTER COLUMN** SALARY decimal (7, 2);

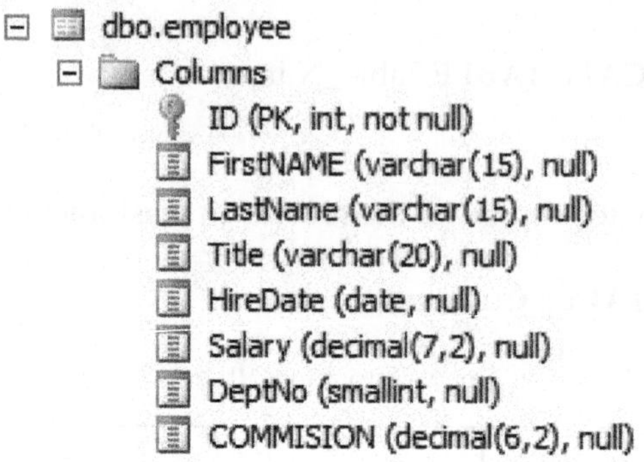

```
⊟ ▦ dbo.employee
   ⊟ 📁 Columns
        🔑  ID (PK, int, not null)
        ▤  FirstNAME (varchar(15), null)
        ▤  LastName (varchar(15), null)
        ▤  Title (varchar(20), null)
        ▤  HireDate (date, null)
        ▤  Salary (decimal(7,2), null)
        ▤  DeptNo (smallint, null)
        ▤  COMMISION (decimal(6,2), null)
```

Figure 5.22 Altering a table in T-SQL

Answer (MySQL):

ALTER TABLE EMPLOYEES **MODIFY** SALARY decimal (7, 2);

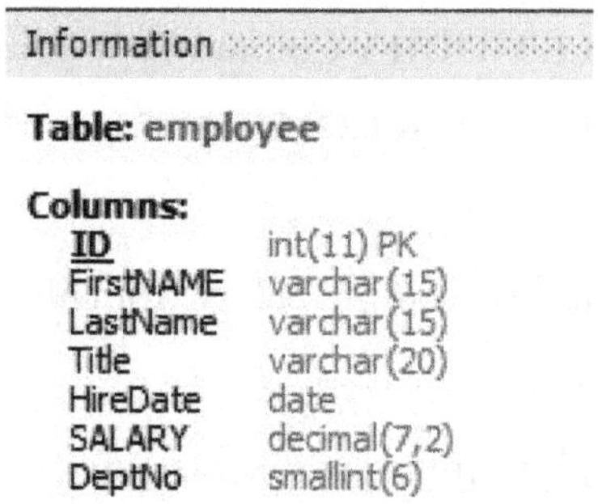

```
Information ░░░░░░░░░░░░░░░░░░░░░░░░░░░░░░░

Table: employee

Columns:
    ID          int(11) PK
    FirstNAME   varchar(15)
    LastName    varchar(15)
    Title       varchar(20)
    HireDate    date
    SALARY      decimal(7,2)
    DeptNo      smallint(6)
```

Figure 5.23 Altering a table in MySQL

Altering a Table (Adding a Column)

Syntax

ALTER TABLE table_name **ADD** Column Type

Question: Write a query to add Commission column to the Employee table.

Answer (Oracle):

ALTER TABLE EMPLOYEES **ADD** COMMISSION NUMBER (6, 2);

COLUMN_NAME	DATA_TYPE
EMPLOYEE_ID	NUMBER(6,0)
FIRST_NAME	VARCHAR2(20 BYTE)
LAST_NAME	VARCHAR2(25 BYTE)
EMAIL	VARCHAR2(25 BYTE)
PHONE	VARCHAR2(20 BYTE)
HIRE_DATE	DATE
JOB_ID	VARCHAR2(10 BYTE)
SALARY	NUMBER(8,2)
MANAGER_ID	NUMBER(6,0)
DEPT_ID	NUMBER(4,0)
COMMISSION	NUMBER(6,2)

Figure 5.24 Adding a column in Oracle

Answer (T-SQL):

ALTER TABLE EMPLOYEES **ADD** COMMISSION decimal (6, 2);

dbo.Employees
Columns
  EMPLOYEE_ID (numeric(6,0), not null)
  FIRST_NAME (varchar(20), null)
  LAST_NAME (varchar(25), null)
  EMAIL (varchar(25), null)
  PHONE (varchar(20), null)
  HIRE_DATE (date, null)
  JOB_ID (varchar(10), null)
  SALARY (decimal(8,2), null)
  MANAGER_ID (numeric(6,0), null)
  DEPT_ID (numeric(4,0), null)
  COMMISSION (decimal(6,2), null)

Figure 5.25 Adding a column in T-SQL

Answer (MySQL):

ALTER TABLE EMPLOYEES **ADD** COMMISSION numeric (6, 2);

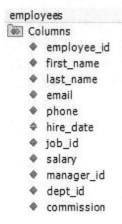

Figure 5.26 Adding a column in MySQL

Summary

Chapter 5 covers the following:

- How to CREATE DATABASE in SQL commands
- How to CREATE DATABASE in GUI tools
- How to CREATE TABLE in SQL commands
- How to CREATE TABLE in GUI tools
- Creating tables that used in this book
- How to RENAME TABLE
- How to DROP DATABASE
- How to TRUNCATE TABLE
- How to ALTER COLUMN

Exercises

5.1

Write SQL commands to create a database.

5.2

Write SQL commands to create three tables in the database.

5.3

Write a query to add a column Budget in the Departments table.

5.4

Write a SQL command to rename table Employees to Emp.

Chapter 6

Data Manipulation
Language (DML)

We have learned Data Definition Language (DDL) in the last chapter. After creating a database and tables the next task is to insert data. If there are errors in a table we should update a record or delete a record. Data Manipulation Language (DML) is used to retrieve and manipulate data in SQL. The main statements for DML are:

> SELECT
> INSERT INTO
> UPDATE... SET
> DELETE FROM

Let us start with INSERT INTO command. How can we select data from a table without data? The following INSERT INTO statements are used to add records to the tables in the last chapter. These statements work for Oracle, T-SQL and MySQL.

INSERT INTO

Description—Inserts a one or more records into a table

Syntax 1

> **INSERT INTO** table (col1, col2, ...)
> VALUES (exp1, exp2, ...);

Syntax 2

> **INSERT INTO** table
> VALUES (exp1, exp2, ...);

Note:

Inserting data in the same order as that in the table for the second style.

Although INSERT INTO statements are the same for Oracle, T-SQL and MySQL but the date format is different:

For example, Oracle date format is **DD-MM-YY**.

> INSERT INTO Employees
> VALUES (100,'Douglas','Grant','DGRANT','650.507.9844','**23-Jan-08**','SH_
> CLERK',2600,114,50);

T-SQL and MySQL date format is **YYYY-MM-DD**.

For example,

> INSERT INTO Employees
>
> VALUES (100,'Douglas','Grant','DGRANT','650.507.9844',**'2008-01-23'**,'SH_CLERK',2600,114,50);

Insert Data into Employees Table (Oracle Date format)

We shall use Oracle SQL Developer to demonstrate inserting data to the Employees table. Enter the following in query worksheet and run the statements.

INSERT INTO Employees

VALUES (100,'Douglas','Grant','DGRANT','650.507.9844','23-Jan-08','SH_CLERK',2600,114,50);

INSERT INTO Employees

VALUES (101,'Adam','Fripp','AFRIPP','650.123.2234','10-Apr-05','SH_MGR',8200,109,50);

INSERT INTO Employees

VALUES (102,'Jennifer','Whalen','JWHALEN','515.123.4444','17-Sep-03','AD_ASST',4400,108,10);

INSERT INTO Employees

VALUES (103,'Michael','Hartstein','MHARTSTE','515.123.5555','17-Feb-04','MK_MGR',13000,109,20);

INSERT INTO Employees

VALUES (104,'Pat','Fay','PFAY','603.123.6666','17-Aug-05','MK_REP',6000,103,20);

INSERT INTO Employees

VALUES (105,'Susan','Mavris','SMAVRIS','515.123.7777','7-Jun-02','HR_MGR',6500,109,40);

INSERT INTO Employees

VALUES (106,'Shelley','Higgins','SHIGGINS','515.123.8080','7-Jun-02','SA_MGR',12008,109,80);

INSERT INTO Employees

VALUES (107,'William','Gietz','WGIETZ','515.123.8181','7-Jun-02','SA_REP',8300,106,80);

INSERT INTO Employees

VALUES (108,'Steven','King','SKING','515.123.4567','17-Jun-03','AD_PRES',24000,,10);

INSERT INTO Employees

VALUES (109,'Lex','De Haan','LDEHAAN','515.123.4569','13-Jan-01','AD_VP',17000,108,10);

INSERT INTO Employees

VALUES (110,'Bruce','Ernst','BERNST','590.423.4568','21-May-07','IT_MGR',6000,109,60);

INSERT INTO Employees

VALUES (111,'Diana','Lorentz','DLORENTZ','590.423.5567','7-Feb-07','IT_ PROG',4200,110,60);

INSERT INTO Employees

VALUES (112,'Nancy','Greenberg','NGREENBE','515.124.4569','17-Aug-02','FI_ MGR',12008,109,90);

INSERT INTO Employees

VALUES (113,'Daniel','Faviet','DFAVIET','515.124.4169','16-Aug-02','FI_ CLERK',3000,112,90);

- Expand Table in the Connection pane and double click the Employees table:

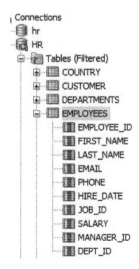

Figure 6.1 Columns in Employees table

- Click **Data** tab to see the data from the Employees table.

EMPLOYEE_ID	FIRST_NAME	LAST_NAME	EMAIL	PHONE	HIRE_DATE	JOB_ID	SALARY	MANAGER_ID	DEPT_ID
100	Douglas	Grant	DGRANT	650.507.9844	23-JAN-08	SH_CLERK	2600	101	50
101	Adam	Fripp	AFRIPP	650.123.2234	10-APR-15	SH_MGR	8200	109	50
102	Jennifer	Whalen	JWHALEN	515.123.4444	17-SEP-13	AD_ASST	4400	108	10
103	Michael	Hartstein	MHARTSTE	515.123.5555	17-FEB-14	MK_MGR	13000	109	20
104	Pat	Fay	PFAY	603.123.6666	17-AUG-15	MK_REP	6000	103	20
105	Susan	Mavris	SMAVRIS	515.123.7777	07-JUN-12	HR_MGR	6500	109	40
106	Shelley	Higgins	SHIGGINS	515.123.8080	07-JUN-12	SA_MGR	12008	109	80
107	William	Gietz	WGIETZ	515.123.8181	07-JUN-12	SA_REP	8300	106	80
108	Steven	King	SKING	515.123.4567	17-JUN-13	AD_PRES	24000	108	10
109	Lex	De Haan	LDEHAAN	515.123.4569	13-JAN-11	AD_VP	17000	108	10
110	Bruce	Ernst	BERNST	590.423.4568	21-FEB-17	IT_MGR	6000	109	60
111	Diana	Lorentz	DLORENTZ	590.423.5567	07-FEB-17	IT_PROG	4200	110	60
112	Nancy	Greenberg	NGREENBE	515.124.4569	17-AUG-12	FI_MGR	12008	109	90
113	Daniel	Faviet	DFAVIET	515.124.4169	16-AUG-12	FI_CLERK	3000	112	90

Figure 6.2 Data in Employees table

Insert Data into Departments Table

INSERT INTO DEPARTMENTS VALUES (10,'Administration',200,1700);
INSERT INTO DEPARTMENTS VALUES (20,'Marketing',201,1800);
INSERT INTO DEPARTMENTS VALUES (30,'Purchasing',114,1700);

INSERT INTO DEPARTMENTS VALUES (40,'Human Resources',203,2400);
INSERT INTO DEPARTMENTS VALUES (50,'Shipping',121,1500);
INSERT INTO DEPARTMENTS VALUES (60,'IT',103,1400);

INSERT INTO DEPARTMENTS VALUES (70,'Public Relations',204,2700);
INSERT INTO DEPARTMENTS VALUES (80,'Sales',145,2500);
INSERT INTO DEPARTMENTS VALUES (90,'Accounting',205,1700);
INSERT INTO DEPARTMENTS VALUES (100,'Customer Service',203,2400);

DEPT_ID	DEPT_NAME	MANAGER_ID	LOCATION_ID
10	Administration	200	1700
20	Marketing	201	1800
30	Purchasing	114	1700
40	Human Resources	203	2400
50	Shipping	121	1500
60	IT	103	1400
70	Public Relations	204	2700
80	Sales	145	2500
90	Accounting	205	1700
100	Customer Service	203	2400
11	Warehousing	114	2400

Figure 6.3 Departments table

Insert Data into Country Table

INSERT INTO COUNTRY VALUES ('AR','Argentina',2);
INSERT INTO COUNTRY VALUES ('AU','Australia',3);
INSERT INTO COUNTRY VALUES ('BE','Belgium',1);

INSERT INTO COUNTRY VALUES ('BR','Brazil',2);
INSERT INTO COUNTRY VALUES ('CA','Canada',2);
INSERT INTO COUNTRY VALUES ('CH','Switzerland',1);
INSERT INTO COUNTRY VALUES ('CN','China',3);
INSERT INTO COUNTRY VALUES ('DE','Germany',1);

INSERT INTO COUNTRY VALUES ('DK','Denmark',1);
INSERT INTO COUNTRY VALUES ('EG','Egypt',4);

INSERT INTO COUNTRY VALUES ('FR','France',1);
INSERT INTO COUNTRY VALUES ('IL','Israel',4);

INSERT INTO COUNTRY VALUES ('IN','India',3);
INSERT INTO COUNTRY VALUES ('IT','Italy',1);
INSERT INTO COUNTRY VALUES ('JP','Japan',3);
INSERT INTO COUNTRY VALUES ('KW','Kuwait',4);
INSERT INTO COUNTRY VALUES ('ML','Malaysia',3);
INSERT INTO COUNTRY VALUES ('MX','Mexico',2);
INSERT INTO COUNTRY VALUES ('NG','Nigeria',4);
INSERT INTO COUNTRY VALUES ('NL','Netherlands',1);
INSERT INTO COUNTRY VALUES ('SG','Singapore',3);
INSERT INTO COUNTRY VALUES ('UK','United Kingdom',1);
INSERT INTO COUNTRY VALUES ('US','United States of America',2);
INSERT INTO COUNTRY VALUES ('ZM','Zambia',4);
INSERT INTO COUNTRY VALUES ('ZW','Zimbabwe',4);

COUNTRY_ID	COUNTRY_NAME	REGION_ID
AR	Argentina	2
AU	Australia	3
BE	Belgium	1
BR	Brazil	2
CA	Canada	2
CH	Switzerland	1
CN	China	3
DE	Germany	1
DK	Denmark	1
EG	Egypt	4
FR	France	1
IL	Israel	4
IN	India	3
IT	Italy	1
JP	Japan	3
KW	Kuwait	4
ML	Malaysia	3
MX	Mexico	2
NG	Nigeria	4
NL	Netherlands	1
SG	Singapore	3
UK	United Kingdom	1
US	United States of ...	2
ZM	Zambia	4

Figure 6.4 Country table

Insert Data into Job Table

INSERT INTO Job VALUES ('AD_PRES','CEO',9000,20000);
INSERT INTO Job VALUES ('AD_VP','VICE President',8000,18000);
INSERT INTO Job VALUES ('AD_ASST','Admin Assistant',5000,6000);
INSERT INTO Job VALUES ('FI_CLERK','Finance Clerk',3000,4000);
INSERT INTO Job VALUES ('FI_MGR','Finance Manager',4000,5000);

INSERT INTO Job VALUES ('SA_REP','Sales Representative',3000,4000);
INSERT INTO Job VALUES ('SA_MGR','Sales Manager',4000,5000);
INSERT INTO Job VALUES ('SH_CLERK','Shipping Clerk',2500,4000);
INSERT INTO Job VALUES ('SH_MGR','Shipping Manager',4000,5000);
INSERT INTO Job VALUES ('IT_PROG','Programmer',4000,5500);
INSERT INTO Job VALUES ('IT_MGR', 'IT Manager', 5000, 6000);
INSERT INTO Job VALUES ('MK_CLERK','Marketing Clerk',3000,4000);
INSERT INTO Job VALUES ('MK_MGR','Marketing Manager',4000,5000);
INSERT INTO Job VALUES ('HR_MGR','Human Resource Manager',4000,5000);

Job Table

JOB_ID	JOB_TITLE	MIN_SALARY	MAX_SALARY
AD_PRES	CEO	9000	25000
AD_VP	VICE President	8000	18000
AD_ASST	Admin Assistant	5000	6000
FI_CLERK	Finance Clerk	3000	4000
FI_MGR	Finance Manager	4000	5000
SA_REP	Sales Representative	3000	4000
SA_MGR	Sales Manager	4000	5000
SH_CLERK	Shipping Clerk	2500	4000
SH_MGR	Shipping Manager	4000	5000
IT_PROG	Programmer	4000	5500
IT_MGR	IT Manager	5000	6000
MK_CLERK	Marketing Clerk	3000	4000
MK_MGR	Marketing Manager	4000	5000
HR_MGR	Human Resource ...	4000	5000

Figure 6.5 Job table

Insert Data into Location Table

INSERT INTO Locations
VALUES (1300,'9450 Kamiya-cho','6823','Hiroshima','','JP');

INSERT INTO Locations
VALUES (1400,'2014 Jabberwocky Rd','26192','Southlake','Texas','US');

INSERT INTO Locations
VALUES (1500,'2011 Interiors Blvd','99236','South San Francisco','California','US');

INSERT INTO Locations
VALUES (1600,'2007 Zagora St','50090','South Brunswick','New Jersey','US');

INSERT INTO Locations
VALUES (1700,'2004 Charade Rd','98199','Seattle','Washington','US');

INSERT INTO Locations
VALUES (1800,'147 Spadina Ave','M5V 2L7','Toronto','Ontario','CA');

INSERT INTO Locations
VALUES (1900,'6092 Boxwood St','YSW 9T2','Whitehorse','Yukon','CA');

INSERT INTO Locations
VALUES (2000,'40-5-12 Laogianggen','190518','Beijing','','CN');

INSERT INTO Locations
VALUES (2200,'12-98 Victoria Street','2901','Sydney','New South Wales','AU');

INSERT INTO Locations
VALUES (2400,'8204 Arthur St','','London','','UK');

INSERT INTO Locations
VALUES (2500,'32 Peachtree Rd','30303','Atlanta','GA','US');

INSERT INTO Locations
VALUES (2700,'560 Main St','37024','Nashville','TN','US');

LOCATION_ID	ADDRESS	POSTAL_CODE	CITY	STATE_PROVINCE	COUNTRY_ID
1300	9450 Kamiya-cho	6823	Hiroshima		JP
1400	2014 Jabberwocky Rd	26192	Southlake	Texas	US
1500	2011 Interiors Blvd	99236	South San Francisco	California	US
1600	2007 Zagora St	50090	South Brunswick	New Jersey	US
1700	2004 Charade Rd	98199	Seattle	Washington	US
1800	147 Spadina Ave	M5V 2L7	Toronto	Ontario	CA
1900	6092 Boxwood St	YSW 9T2	Whitehorse	Yukon	CA
2000	40-5-12 Laogianggen	190518	Beijing		CN
2200	12-98 Victoria Street	2901	Sydney	New South Wales	AU
2400	8204 Arthur St		London		UK
2500	32 Peachtree Rd	30303	Atlanta	GA	US
2700	560 Main St	37024	Nashville	TN	US

Figure 6.6 Locations table

Insert Data into Regions Table

INSERT INTO REGIONS VALUES (1,'Europe');
INSERT INTO REGIONS VALUES (2,'Americas');
INSERT INTO REGIONS VALUES (3,'Asia');
INSERT INTO REGIONS VALUES (4,'Middle East and Africa');

Regions Table

REGION_ID	REGION_NAME
1	Europe
2	Americas
3	Asia
4	Middle East and Africa

Figure 6.7 Regions table

SELECT Statement

Description—Retrieve records from one or more tables

Syntax   **SELECT** column(s)
FROM tables
[WHERE conditions] —Optional
[ORDER BY column(s) [ASC | DESC]]; —Optional

SELECT All Columns

Syntax   **SELECT** *
FROM tables
[WHERE conditions] —Optional
[ORDER BY column(s) [ASC | DESC]]; —Optional

Note: ASC – Ascending order
DESC – Descending order

Question 1: Write a query to select all the data from the Departments table.

Answer:

SELECT *
FROM Departments;

Oracle SQL				T-SQL				MySQL			
DEPT_ID	DEPT_NAME	MANAGER_ID	LOCATION_ID	DEPT_ID	DEPT_NAME	MANAGER_ID	LOCATION_ID	dept_id	dept_name	manager_id	location_id
10	Administration	200	1700	10	Administration	200	1700	10	Administration	200	1700
20	Marketing	201	1800	20	Marketing	201	1800	20	Marketing	201	1800
30	Purchasing	114	1700	30	Purchasing	114	1700	30	Purchasing	114	1700
40	Human Resources	203	2400	40	Human Resources	203	2400	40	Human Resources	203	2400
50	Shipping	121	1500	50	Shipping	121	1500	50	Shipping	121	1500
60	IT	103	1400	60	IT	103	1400	60	IT	103	1400
70	Public Relations	204	2700	70	Public Relations	204	2700	70	Public Relations	204	2700
80	Sales	145	2500	80	Sales	145	2500	80	Sales	145	2500
90	Accounting	205	1700	90	Accounting	205	1700	90	Accounting	205	1700
100	Customer Service	203	2400	100	Customer Service	203	2400	100	Customer Service	203	2400

Figure 6.8 Query output for question 1

SELECT Specific Column(s)

Question 2: Write a query to display department names.

Answer: **SELECT** dept_name
                 FROM Departments;

Oracle SQL	T-SQL	MySQL
DEPT_NAME	DEPT_NAME	dept_name
Administration	Administration	Administration
Marketing	Marketing	Marketing
Purchasing	Purchasing	Purchasing
Human Resources	Human Resources	Human Resources
Shipping	Shipping	Shipping
IT	IT	IT
Public Relations	Public Relations	Public Relations
Sales	Sales	Sales
Accounting	Accounting	Accounting
Customer Service	Customer Service	Customer Service

Figure 6.9 Query output for question 2

DISTINCT Clause

Description: Eliminates duplicates from the result of a SELECT statement.

Syntax

          SELECT **DISTINCT** column_name
          FROM table_name;

Question 3: Write a query to select the minimal salary without duplicates.

Answer:

          SELECT **DISTINCT** Min_Salary
          FROM Job;

Oracle SQL	T-SQL	MySQL
MIN_SALARY	Min_Salary	Min_Salary
2500	2500	2500
3000	3000	3000
4000	4000	4000
5000	5000	5000
8000	8000	8000
9000	9000	9000

Figure 6.10 Query output for question 3

WHERE Clause

Description: When the condition is true the WHERE clause filters unwanted rows from the result.

Syntax   SELECT column(s)
                    FROM table
                    WHERE conditions;

Question 4: Write a query to get the name of a country with country_id "IT".

Answer:
                    SELECT country_name
                    FROM COUNTRY
                    WHERE country_id = 'IT';

Oracle SQL	T-SQL	MySQL
COUNTRY_NAME	country_name	country_name
Italy	Italy	Italy

Figure 6.11 Query output for question 4

Arithmetic Operators

You can create an expression with number and field value using arithmetic operators: Addition (+), Subtraction (−), Multiplication (*), Division (/).

Question 5: Write a query to display job title, minimal salary with 10% increased minimal salary.

Answer (Oracle):

SELECT Job_Title, Min_Salary, Min_Salary * 1.1
FROM JOB;

JOB_TITLE	MIN_SALARY	MIN_SALARY*1.1
CEO	9000	9900
VICE President	8000	8800
Admin Assistant	5000	5500
Finance Clerk	3000	3300
Finance Manager	4000	4400
Sales Representative	3000	3300
Sales Manager	4000	4400
Shipping Clerk	2500	2750
Shipping Manager	4000	4400
Programmer	4000	4400
IT Manager	5000	5500
Marketing Clerk	3000	3300
Marketing Manager	4000	4400
Human Resource Manager	4000	4400

Figure 6.12 Query output for question 5

Answer (T-SQL & MySQL):

SELECT Job_Title, Min_Salary, Min_Salary * 1.1 AS 'Min_Salary * 1.1'
FROM JOBS;

Job_Title	Min_Salary	Min_Salary * 1.1	Job_Title	Min_Salary	Min_Salary * 1.1
CEO	9000	9900.0	CEO	9000	9900.0
VICE President	8000	8800.0	VICE President	8000	8800.0
Admin Assistant	5000	5500.0	Admin Assistant	5000	5500.0
Finance Clerk	3000	3300.0	Finance Clerk	3000	3300.0
Finance Manager	4000	4400.0	Finance Manager	4000	4400.0
Sales Representative	3000	3300.0	Sales Representative	3000	3300.0
Sales Manager	4000	4400.0	Sales Manager	4000	4400.0
Shipping Clerk	2500	2750.0	Shipping Clerk	2500	2750.0
Shipping Manager	4000	4400.0	Shipping Manager	4000	4400.0
Programmer	4000	4400.0	Programmer	4000	4400.0
IT Manager	5000	5500.0	IT Manager	5000	5500.0
Marketing Clerk	3000	3300.0	Marketing Clerk	3000	3300.0
Marketing Manager	4000	4400.0	Marketing Manager	4000	4400.0
Human Resource Manager	4000	4400.0	Human Resource Manager	4000	4400.0

Figure 6.13 Query output for question 5 (Left: T-SQL, Right: MySQL)

Order of Arithmetic Operators

Priority Level 1:   % Modulo   / Division   * Multiplication

Priority Level 2:   + Addition   – Minus

General Rule

Modulo operator, multiplication operator and division operator are calculated first then addition and minus operator are processed.

In order to change the priority we can add parentheses.

Comparison Operators

=	Equal to
>	Greater than
<	Less than
>=	Greater than equal to
<=	Less than equal to
<>	Not equal to

Question 6: Write a query to get employee names with hire date greater than Jan 1st, 2017.

Answer (Oracle):

> SELECT First_Name, Last_Name, hire_date
> FROM Employees
> WHERE hire_date > '01-JAN-17';

Answer (T-SQL & MySQL):

> SELECT First_Name, Last_Name, hire_date
> FROM Employees
> WHERE hire_date > '2017-01-01';

Oracle SQL			T-SQL			MySQL		
FIRST_NAME	LAST_NAME	HIRE_DATE	First_Name	Last_Name	hire_date	First_Name	Last_Name	hire_date
Bruce	Ernst	21-MAY-17	Bruce	Ernst	2017-02-21	Bruce	Ernst	2017-02-21
Diana	Lorentz	07-FEB-17	Diana	Lorentz	2017-01-18	Diana	Lorentz	2017-01-18

Figure 6.14 Query output for question 6

AND Condition

Description: Test for two or more conditions in a SELECT, INSERT, UPDATE, or DELETE statement. All conditions must be true for a record to be selected.

Syntax

> SELECT column(s)
> FROM table
> WHERE condition {**AND** condition};

Question 7: Write a query to get employee names with hire date greater than January 1st, 2017 and salary less than $5,000.

Answer (Oracle):

> SELECT First_Name, Last_Name, hire_date
> FROM Employees
> WHERE hire_date > '01-JAN-17'
> **AND** salary < 5000;

Answer (T-SQL & MySQL):

> SELECT First_Name, Last_Name, hire_date
> FROM Employees
> WHERE hire_date > '2017-01-01'
> **AND** salary < 5000;

Oracle SQL			T-SQL			MySQL		
FIRST_NAME	LAST_NAME	HIRE_DATE	First_Name	Last_Name	hire_date	First_Name	Last_Name	hire_date
Diana	Lorentz	07-FEB-17	Diana	Lorentz	2017-01-18	Diana	Lorentz	2017-01-18

Figure 6.15 Query output for question 7

OR Condition

Description: Test multiple conditions in a SELECT, INSERT, UPDATE, or DELETE statement. Any one of the conditions must be true for a record to be selected.

Syntax

> SELECT column(s)
> FROM table_name
> WHERE condition {**OR** condition};

Question 8: Write a query to get employee job title for Shipping Manager or minimal salary is $5,000.

Answer: SELECT Job_Title, Min_Salary
               FROM JOB
               WHERE Job_Title = 'Shipping Manager'
               OR Min_Salary = 5000;

Oracle SQL		T-SQL		MySQL	
JOB_TITLE	MIN_SALARY	Job_Title	Min_Salary	Job_Title	Min_Salary
Admin Assistant	5000	Admin Assistant	5000	Admin Assistant	5000
Shipping Manager	4000	Shipping Manager	4000	Shipping Manager	4000
IT Manager	5000	IT Manager	5000	IT Manager	5000

Figure 6.16 Query output for question 8

IN Condition

Description: Test if an expression matches any value in a list of VALUES. It can reduce the need for multiple OR conditions in a SELECT, INSERT, UPDATE, or DELETE statement.

Syntax   SELECT column(s)
               FROM table
               WHERE column_name **IN** (value1, value2,...);
               [WHERE column_name **NOT IN** (value1, value2,...);]

Question 9: Write a query to state/province and country ID with city in Seattle or Toronto.

Answer:
               SELECT City, State_Province, Country_ID
               FROM LOCATIONS
               WHERE city **IN** ('Seattle', 'Toronto');

Oracle SQL			T-SQL			MySQL		
CITY	STATE_PROVINCE	COUNTRY_ID	City	State_Province	Country_ID	City	State_Province	Country_ID
Seattle	Washington	US	Seattle	Washington	US	Seattle	Washington	US
Toronto	Ontario	CA	Toronto	Ontario	CA	Toronto	Ontario	CA

Figure 6.17 Query output for question 9

BETWEEN Condition

Description: To check if an expression is within a range of VALUES.

Syntax
SELECT column(s)
FROM table_name
WHERE column_name **BETWEEN** value1 **AND** value2;

Question 10: Write a query to get employee names with hire date from January 1st, 2014 to December 31, 2015.

Answer (Oracle):

SELECT First_Name, Last_Name, hire_date
FROM Employees
WHERE hire_date **BETWEEN** '01-JAN-14' **AND** '31-DEC-15';

Answer (T-SQL & MySQL):

SELECT First_Name, Last_Name, hire_date
FROM Employees
WHERE hire_date **BETWEEN** '2014-01-01' **AND** '2015-12-31';

Oracle SQL			T-SQL			MySQL		
FIRST_NAME	LAST_NAME	HIRE_DATE	First_Name	Last_Name	hire_date	First_Name	Last_Name	hire_date
Adam	Fripp	10-APR-15	Adam	Fripp	2015-04-10	Adam	Fripp	2015-04-10
Michael	Hartstein	17-FEB-14	Michael	Hartstein	2014-02-17	Michael	Hartstein	2014-02-17
Pat	Fay	17-AUG-15						

Figure 6.18 Query output for question 10

IS NULL

Description: Uses IS NULL to test a NULL value.

Syntax (Oracle)   expression **IS NULL**

Syntax (T-SQL & MySQL)   expression = ' '

Question 11: Write a query to find cities without states or provinces.

Answer (Oracle):

SELECT CITY, STATE_PROVINCE, COUNTRY_ID
FROM Locations
WHERE STATE_PROVINCE **IS NULL**;

Answer (T-SQL & MySQL):

> SELECT CITY, STATE_PROVINCE, COUNTRY_ID
> FROM Locations
> WHERE STATE_PROVINCE = ' ';

Oracle SQL			T-SQL			MySQL		
CITY	STATE_PROVINCE	COUNTRY_ID	CITY	STATE_PROVINCE	COUNTRY_ID	city	state_province	country_id
Hiroshima	(null)	JP	Hiroshima		JP	Hiroshima		JP
Beijing	(null)	CN	Beijing		CN	Beijing		CN
London	(null)	UK	London		UK	London		UK

Figure 6.19 Query output for question 11

IS NOT NULL

Description: Uses IS NULL to test a NOT NULL value.

Syntax (Oracle)   expression **IS NOT NULL**

Syntax (T-SQL & MySQL)   expression < > ' '

Question 12: Write a query to find cities with states or provinces.

Answer (Oracle):

> SELECT CITY, STATE_PROVINCE, COUNTRY_ID
> FROM Locations
> WHERE STATE_PROVINCE **IS NOT NULL**;

Answer (T-SQL & MySQL):

> SELECT CITY, STATE_PROVINCE, COUNTRY_ID
> FROM Locations
> WHERE STATE_PROVINCE <> ' ';

CITY	STATE_PROVINCE	COUNTRY_ID
Southlake	Texas	US
South San Francisco	California	US
South Brunswick	New Jersey	US
Seattle	Washington	US
Toronto	Ontario	CA
Whitehorse	Yukon	CA
Sydney	New South Wales	AU
Nashville	TN	US
Atlanta	GA	US

Figure 6.20 Query output for question 11 (Oracle)

CITY	STATE_PROVINCE	COUNTRY_ID
Southlake	Texas	US
South San Francisco	California	US
South Brunswick	New Jersey	US
Seattle	Washington	US
Toronto	Ontario	CA
Whitehorse	Yukon	CA
Sydney	New South Wales	AU
Atlanta	GA	US
Nashville	TN	US

city	state_province	country_id
Southlake	Texas	US
South San Francisco	California	US
South Brunswick	New Jersey	US
Seattle	Washington	US
Toronto	Ontario	CA
Whitehorse	Yukon	CA
Sydney	New South Wales	AU
Atlanta	GA	US
Nashville	TN	US

Figure 6.21 Query output for question 12 (Left: T-SQL, Right: MySQL)

LIKE Condition

Description: Uses wildcards to perform pattern matching in a query.

% (percent sign)—represents zero, one, or more characters.

_ (underscore)—represents exactly one character.

Syntax

SELECT column(s)

FROM table_name

WHERE expression **LIKE** pattern

Question 13: Write a query to get country IDs and country names that begins with "U".

Answer:

SELECT country_id, country_name

FROM COUNTRY

WHERE country_name

LIKE 'U%';

Oracle SQL		T-SQL		MySQL	
COUNTRY_ID	COUNTRY_NAME	country_id	country_name	country_id	country_name
UK	United Kingdom	UK	United Kingdom	UK	United Kingdom
US	United States of America	US	United States of America	US	United States of America

Figure 6.22 Query output for question 13

Question 14: Write a query to get country ID and full country name for "Isr?el".

Answer:

SELECT country_id, country_name

FROM COUNTRY

WHERE country_name
LIKE 'Isr_el';

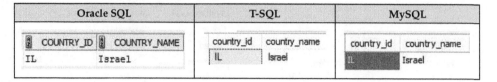

Figure 6.23 Query output for question 14

ORDER BY Clause

Description: To sort the records in the result set for a SELECT statement.

Syntax

SELECT Column(s)
FROM tables
WHERE conditions
ORDER BY expression [DESC];

Note: **Order By** express returns result in ascending order by default
DESC—descending order

Question 15: Write a query to display the department names in ascending order.

Answer:

SELECT Dept_Name
FROM Departments;
ORDER BY Dept_Name;

Oracle SQL	T-SQL	MySQL
DEPT_NAME	Dept_Name	Dept_Name
Accounting	Accounting	Accounting
Administration	Administration	Administration
Customer Service	Customer Service	Customer Service
Human Resources	Human Resources	Human Resources
IT	IT	IT
Marketing	Marketing	Marketing
Public Relations	Public Relations	Public Relations
Purchasing	Purchasing	Purchasing
Sales	Sales	Sales
Shipping	Shipping	Shipping

Figure 6.24 Query output for question 15

Question 16: Write a query to display the department names in descending order.

Answer:

SELECT Dept_Name
FROM Departments
ORDER BY Dept_Name **DESC;**

Oracle SQL	T-SQL	MySQL
🔖 DEPT_NAME	Dept_Name	Dept_Name
Shipping	Shipping	Shipping
Sales	Sales	Sales
Purchasing	Purchasing	Purchasing
Public Relations	Public Relations	Public Relations
Marketing	Marketing	Marketing
IT	IT	IT
Human Resources	Human Resources	Human Resources
Customer Service	Customer Service	Customer Service
Administration	Administration	Administration
Accounting	Accounting	Accounting

Figure 6.25 Query output for question 16

Question 17: Write a query to display employee names, salary and department ID by ascending order of department and salary.

Answer (Oracle):

SELECT first_name | | ' ' | | last_name AS Full_Name, Salary, Dept_ID
FROM employees
ORDER BY Dept_ID, Salary;

Answer (T-SQL):

SELECT first_name + ' ' + last_name AS Full_Name, Salary, Dept_ID
FROM employees
ORDER BY Dept_ID, Salary;

Answer (MySQL):

SELECT CONCAT(first_name, ' ', last_name) AS Full_Name, Salary, Dept_ID
FROM employees
ORDER BY Dept_ID, Salary;

Oracle SQL			T-SQL			MySQL		
FULL_NAME	SALARY	DEPT_ID	Full_Name	Salary	Dept_ID	Full_Name	Salary	Dept_ID
Jennifer Whalen	4400	10	Jennifer Whalen	4400.00	10	Jennifer Whalen	4400.00	10
Lex De Haan	17000	10	Lex De Haan	17000.00	10	Lex De Haan	17000.00	10
Steven King	24000	10	Steven King	24000.00	10	Steven King	24000.00	10
Pat Fay	6000	20	Pat Fay	6000.00	20	Pat Fay	6000.00	20
Michael Hartstein	13000	20	Michael Hartstein	13000.00	20	Michael Hartstein	13000.00	20
Susan Mavris	6500	40	Susan Mavris	6500.00	40	Susan Mavris	6500.00	40
Douglas Grant	2600	50	Douglas Grant	2600.00	50	Douglas Grant	2600.00	50
Adam Fripp	8200	50	James Fripp	8200.00	50	James Fripp	8200.00	50
Diana Lorentz	4200	60	Diana Lorentz	4200.00	60	Diana Lorentz	4200.00	60
Bruce Ernst	6000	60	Bruce Ernst	6000.00	60	Bruce Ernst	6000.00	60
William Gietz	8300	80	William Gietz	8300.00	80	William Gietz	8300.00	80
Shelley Higgins	12008	80	Shelley Higgins	12008.00	80			
Daniel Faviet	3000	90	Daniel Faviet	3000.00	90			
Nancy Greenberg	12008	90	Nancy Greenberg	12008.00	90			

Figure 6.26 Query output for question 17

Using Aliases

Description: Creates a temporary name for columns or tables.

Syntax (Column Aliases)   SELECT column_name **AS** alias_name
FROM Tables

Syntax (Table Aliases)   SELECT column_name
FROM Tables **AS** alias_name

Question 18: Write a query to use alias names for minimum salary and maximum salary.

Answer:
SELECT min(salary) **AS** Minimum_Salary,
max(salary) **AS** Maximum_Salary
FROM Employees;

Oracle SQL		T-SQL		MySQL	
MINIMUM_SALARY	MAXIMUM_SALARY	Minimum_Salary	Maximum_Salary	Minimum_Salary	Maximum_Salary
2600	24000	2600.00	24000.00	2600.00	24000.00

Figure 6.27 Query output for question 18

Note:
See Chapter 7 for min( ) and max( ) functions. For table aliases examples see Chapter 10.

INSERT multiple records from an Existing table

Syntax   **INSERT INTO** table (col1, col2, ...)
                      SELECT col1, col2, ...
                      FROM source_tables
                      [WHERE conditions];

We use Oracle SQL as example. Suppose we have following records in Customers table:

ID	FIRSTNAME	LASTNAME	CITY	COUNTRY	PHONE
1	Howard	Bell	Atlanta	USA	(678) 555-7629
2	Alice	Carter	Boston	USA	(617) 213-6874
3	Carine	Schmitt	Nantes	France	40.32.21.21
4	Paolo	Accorti	Torino	Italy	011-4988260
5	Helen	Bennett	Barcelona	Spain	(93) 203 4560

Figure 6.28 Customers table

Question 19: Write a query to insert all the records in Customers table to the Employees table.

Answer:
                      INSERT INTO Employees (Employee_ID, First_Name, Last_Name, phone)
                      SELECT ID, FirstName, LastName, phone
                      FROM Customers;

EMPLOYEE_ID	FIRST_NAME	LAST_NAME	EMAIL	PHONE
1	Howard	Bell	(null)	(678) 555-7629
2	Alice	Carter	(null)	(617) 213-6874
3	Carine	Schmitt	(null)	40.32.21.21
4	Paolo	Accorti	(null)	011-4988260
5	Helen	Bennett	(null)	(93) 203 4560
100	Douglas	Grant	DGRANT	650.507.9844
101	Adam	Fripp	AFRIPP	650.123.2234
102	Jennifer	Whalen	JWHALEN	515.123.4444
103	Michael	Hartstein	MHARTSTE	515.123.5555
104	Pat	Fay	PFAY	603.123.6666
105	Susan	Mavris	SMAVRIS	515.123.7777

Figure 6.29 Query output for question 19

UPDATE Statement

Description: Updates existing records in the tables

Syntax   **UPDATE** table
                SET col1 = value1, col2 = value2, ...
                [WHERE conditions];

Question 20: Write a query to update the first name "Adam" to "James" (Employee ID is 101).

Answer:
                UPDATE Employees
                SET First_Name = 'James'
                WHERE Employee_Id = 101;

To check the updated record:

                SELECT First_Name, Last_Name
                FROM Employees
                WHERE Employee_id = 101;

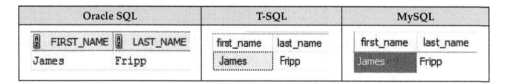

Oracle SQL		T-SQL		MySQL	
FIRST_NAME	LAST_NAME	first_name	last_name	first_name	last_name
James	Fripp	James	Fripp	James	Fripp

Figure 6.30 Query output for question 20

DELETE Statement

Description: Deletes one or more records from a table.

Syntax   **DELETE FROM** table
                [WHERE conditions];

Question 21: Write a query to remove department Warehouse.

Answer:
                DELETE FROM Departments
                WHERE Dept_Id = 11;

Data Control Language (DCL)

We mentioned in Chapter 5 that there are two main statements in Data Control Language: GRANT and REVOKE. Actually, we have already used GRANT statement on page 48 when we create database in Oracle.

GRANT: Assigns privileges on database objects to a user. The system privileges can be CONNECT, CREATE. The table privileges can be INSERT, UPDATE, DELETE, or ALTER.

Syntax

> **GRANT** system privileges **TO** user;
> **GRANT** table privileges ON objects TO user;

In the following example, we GRANT CONNECT and DBA privileges to a user TEST_DB.
> **GRANT** CONNECT, DBA TO TEST_DB;

In the following example, we GRANT INSERT privileges to a user TEST_DB.
> **GRANT** INSERT ON Departments TO TEST_DB;

If you want to grant all the privileges to a user you can use ALL keyword:
> **GRANT** ALL ON Departments TO TEST_DB;

REVOKE: Removes privileges from a user.

Syntax

> **REVOKE** privileges ON objects FROM user;

Example:

> **REVOKE** DBA TO TEST_DB;
> **REVOKE** INSERT ON Departments TO TEST_DB;

Summary

Chapter 6 covers the following:

- Data Manipulation Language (DML)
- Using INSERT INTO statements to insert data to the six tables used in this book
- SELECT statements in Oracle SQL, T-SQL and MySQL
- Using arithmetic operators
- Using comparison operators
- UPDATE and DELETE statements

Exercises

6.1

Write a query to display all the countries.

6.2

Write a query to display specific columns like email and phone number for all the employees.

6.3

Write a query to display the data of employee whose last name is "Fay".

6.4

Write a query to find the hire date for employees whose last name is "Grant" or "Whalen".

6.5

Write a query to display name of the employee who is shipping manager.

6.6

Write a query to get all the employees who work for department 20.

6.7

Write a query to display the departments in the descending order.

6.8

Write a query to display all the employees whose last name starts with "M".

6.9

Display name of the employees whose hire dates are between 2015 and 2017.

6.10

Write a query to display jobs where the maximum salary is less than 5000.

6.11

Write a query to display email address in lower case.

6.12

Write a query to display name of the employees who were hired in 2015.

6.13

Write a query to insert an employee "Paul Newton" in department 20.

6.14

Write a query to delete the shipping department.

Chapter 7

Aggregate Functions and GROUP BY Clause

Database developers often need to answer questions such as how many employees are there in each department. In order to write queries for this kind of summary questions we need to understand the aggregate functions and Group By clause. We will use the Employees table for sample data in this chapter.

EMPLOYEE_ID	FIRST_NAME	LAST_NAME	EMAIL	PHONE	HIRE_DATE	JOB_ID	SALARY	MANAGER_ID	DEPT_ID
100 Douglas	Grant	DGRANT	650.507.9844	23-JAN-08	SH_CLERK	2600	101	50
101 Adam	Fripp	AFRIPP	650.123.2234	10-APR-15	SH_MGR	8200	109	50
102 Jennifer	Whalen	JWHALEN	515.123.4444	17-SEP-13	AD_ASST	4400	108	10
103 Michael	Hartstein	MHARTSTE	515.123.5555	17-FEB-14	MK_MGR	13000	109	20
104 Pat	Fay	PFAY	603.123.6666	17-AUG-15	MK_REP	6000	103	20
105 Susan	Mavris	SMAVRIS	515.123.7777	07-JUN-12	HR_MGR	6500	109	40
106 Shelley	Higgins	SHIGGINS	515.123.8080	07-JUN-12	SA_MGR	12008	109	80
107 William	Gietz	WGIETZ	515.123.8181	07-JUN-12	SA_REP	8300	106	80
108 Steven	King	SKING	515.123.4567	17-JUN-13	AD_PRES	24000	108	10
109 Lex	De Haan	LDEHAAN	515.123.4569	13-JAN-11	AD_VP	17000	108	10
110 Bruce	Ernst	BERNST	590.423.4568	21-FEB-17	IT_MGR	6000	109	60
111 Diana	Lorentz	DLORENTZ	590.423.5567	07-FEB-17	IT_PROG	4200	110	60
112 Nancy	Greenberg	NGREENBE	515.124.4569	17-AUG-12	FI_MGR	12008	109	90
113 Daniel	Faviet	DFAVIET	515.124.4169	16-AUG-12	FI_CLERK	3000	112	90

Figure 7.1 Employees table

Aggregate Functions

Syntax   **SELECT Aggregate Function** (column_name)
                From Table

Below are the main aggregate functions:

    AVG ( ):   To select the average value for certain table column.

    COUNT ( ):   To count the number of rows in a database table.

MAX ( ) :   To select the highest value for a certain column.
It returns the maximum value for numeric data column.
It returns the latest date for date column.
It returns the last records for a character column.

MIN ( ) :   To select the lowest value for a certain column.
It returns the minimum value for a numeric data column.
It returns the earliest date for a date column.
It returns the first records for a character column.

SUM ( ) :   To select the total for a numeric column.

ROUND ( ) :   To round a number to a specified decimal places.

AVG ( ) Function

Question: Write a query to find average salary in the Employees table.

Answer (Oracle & MySQL):

SELECT **AVG**(salary)
FROM Employees;

Answer (T-SQL):

SELECT **AVG** (Salary) AS 'Average Salary'
FROM Employees;

Oracle SQL	T-SQL	MySQL
AVG(SALARY) 9086.8571428571	Average Salary 9086.857142	avg(salary) 9086.857143

Figure 7.2 Query output for AVG( ) function

COUNT ( ) Function

Question: Write a query to count the Employees.

Answer (Oracle & MySQL):

SELECT **COUNT**(*)
FROM Employees;

Answer (T-SQL):

>
> SELECT **COUNT** (*) AS Count_of_Employees
> FROM Employees;

Oracle SQL	T-SQL	MySQL
COUNT(*)	Count_of_Employees	count(*)
14	14	14

Figure 7.3 Query output COUNT( ) function

MIN ( ) Function

Question: Write a query to get the minimum salary in the Employees table.

Answer (Oracle & MySQL):

>
> SELECT **MIN**(Salary)
> FROM Employees;

Answer (T-SQL):

>
> SELECT **MIN** (Salary) AS Max_Salary
> FROM Employees;

Oracle SQL	T-SQL	MySQL
MIN(SALARY)	Min_Salary	MIN(Salary)
2600	2600.00	2600.00

Figure 7.4 Query output MIN( ) function (number type)

Question: Write a query to display the first record in the last name column.

Answer (Oracle & MySQL):

>
> SELECT **MIN**(Last_Name)
> FROM Employees;

Answer (T-SQL):

> SELECT **MIN**(Last_Name) AS Last_Name
> FROM Employees;

Oracle SQL	T-SQL	MySQL
MIN(LAST_NAME)	Last_Name	MIN(Last_Name)
De Haan	De Haan	De Haan

Figure 7.5 Query output for MIN( ) function (character type)

MAX ( ) Function

Question: Write a query to get maximum salary in the Employees table.

Answer (Oracle & MySQL):

> SELECT **MAX**(Salary)
> FROM Employees;

Answer (T-SQL):

> SELECT **MAX** (Salary) AS Max_Salary
> FROM Employees;

Oracle SQL	T-SQL	MySQL
MAX(SALARY)	Max_Salary	MAX(Salary)
24000	24000.00	24000.00

Figure 7.6 Query output for MAX( ) function (number type)

Question: Write a query to display the last record in the last name column.

Answer (Oracle & MySQL):

> SELECT **MAX**(Last_Name)
> FROM Employees;

Answer (T-SQL):

> SELECT **MAX** (Last_Name) AS Last_Name
> FROM Employees;

Oracle SQL	T-SQL	MySQL
MAX(LAST_NAME) Whalen	Last_Name Whalen	MAX(Last_Name) Whalen

Figure 7.7 Query output for MAX( ) function (character type)

Question: Write a query to display the latest hire date in the Employees table.

Answer (Oracle & MySQL):

> SELECT **MAX**(Hire_Date)
> FROM Employees;

Answer (T-SQL):

> SELECT **MAX** (Hire_Date) AS Hire_Date
> FROM Employees;

Oracle SQL	T-SQL	MySQL
MAX(HIRE_DATE) 21-FEB-17	Hire_Date 2017-02-21	MAX(Hire_Date) 2017-02-21

Figure 7.8 Query output for MAX( ) function (date type)

SUM ( ) Function

Question: Write a query to calculate the total amount of employee salary from the Employee table.

Answer (Oracle & MySQL):

> SELECT **SUM**(Salary)
> FROM Employees;

Answer (T-SQL):

>   SELECT **SUM** (Salary) AS Total_Salary
>   FROM Employees;

Oracle SQL	T-SQL	MySQL
SUM(SALARY)	Total_Salary	SUM(Salary)
127216	127216.00	127216.00

Figure 7.9 Query output for SUM( ) function

GROUP BY and HAVING Clause

The GROUP BY statement is used with the aggregate functions to group data from a column. HAVING clause is used in a GROUP BY statement. It sets conditions on group(s). HAVING clause is used in SELECT statement.

Syntax   SELECT Aggregate Function (column_name)
FROM tables
[WHERE conditions]
GROUP BY column_name
[HAVING conditions]
[ORDER BY column(s) [ASC I DESC]];

GROUP BY with AVG ( ) Function

Question: Write a query to find average salary for each department.

Answer (Oracle & MySQL):

>   SELECT **AVG** (salary), Dept_ID
>   FROM Employees
>   **GROUP BY** Dept_ID
>   ORDER BY Dept_ID;

Answer (T-SQL):

>   SELECT **AVG** (Salary) AS 'Average Salary', Dept_ID
>   FROM Employees
>   **GROUP BY** Dept_ID
>   ORDER BY Dept_ID;

Oracle SQL			T-SQL		MySQL	
AVG(SALARY)	DEPT_ID		Average Salary	Dept_ID	avg(salary)	Dept_ID
15133.3333...	10		15133.33	10	15133.333333	10
9500	20		9500.00	20	9500.000000	20
6500	40		6500.00	40	6500.000000	40
5400	50		5400.00	50	5400.000000	50
5100	60		5100.00	60	5100.000000	60
10154	80		10154.00	80	10154.000000	80
7504	90		7504.00	90	7504.000000	90

Figure 7.10 Query output for GROUP BY with AVG( ) function

Note:

If you do not list Dept_ID in the SELECT clause the result has only one column Average Salary. It is not clear for which group (department). So always list the Group By column(s) in the SELECT clause.

GROUP BY with COUNT ( ) Function

Question: Write a query to count number of employees in every department.

Answer (Oracle & MySQL):

> SELECT **COUNT**(Employee_ID), Dept_ID
> FROM Employees
> **GROUP BY** Dept_ID
> ORDER BY Dept_ID;

Answer (T-SQL):

> SELECT **COUNT** (Employee_ID) AS 'Nunber of Employees', Dept_ID
> FROM Employees
> **GROUP BY** Dept_ID
> ORDER BY Dept_ID;

Oracle SQL		T-SQL		MySQL	
COUNT(EMPLOYEE_ID)	DEPT_ID	Number of Employees	Dept_ID	count(Employee_ID)	Dept_ID
3	10	3	10	3	10
2	20	2	20	2	20
1	40	1	40	1	40
2	50	2	50	2	50
2	60	2	60	2	60
2	80	2	80	2	80
2	90	2	90	2	90

Figure 7.11 Query output for GROUP BY with COUNT( ) function

GROUP BY with HAVING Example

Question: Write a query to count employees for the departments that have three employees.

Answer (Oracle & MySQL):

> SELECT COUNT (Employee_ID), Dept_ID
> FROM Employees
> **GROUP BY** Dept_ID
> **HAVING** COUNT (Employee_ID) = 3;

Answer (T-SQL):

> SELECT COUNT (Employee_ID) AS 'Nunber of Employees', Dept_ID
> FROM Employees
> **GROUP BY** Dept_ID
> **HAVING** COUNT (Employee_ID) = 3;

Oracle SQL		T-SQL		MySQL	
COUNT(EMPLOYEE_ID)	DEPT_ID	Number of Employees	Dept_ID	COUNT(Employee_ID)	Dept_ID
3	10	3	10	3	10

Figure 7.12 Query output for GROUP BY with HAVING example

Summary

Chapter 7 covers the following:

- Using aggregate function AVG ( ), COUNT ( ), MAX ( ), MIN ( ), SUM ( ) and ROUND ( ).
- Using GROUP BY and HAVING clauses.
- Using GROUP BY with AVG ( ) Function.
- Using GROUP BY with COUNT ( ) Function.

Exercises

7.1

Write a query to display the number of cities in the country.

7.2

Write a query to display minimal salary of employees in every department.

7.3

Write a query to display maximum salary of employees in every department.

7.4

Write a query to display sum of salary of employees in every department.

7.5

Write a query to display sum of salary in every department.

7.6

Display the ID of departments with average salary greater than 15000.

7.7

Write a query to display the number of employees managed by the manager.

7.8

Write a query to display managers who are managing more than 3 employees.

7.9

Write a query to increase salary of employee 111 to 5000.

Chapter 8

Functions

Common Number Functions

The numeric functions take a numeric input as an expression and return numeric values. The return type for most of the numeric functions is NUMBER.

For aggregate functions **AVG ( )**, **COUNT ( )**, **MAX ( )**, **MIN ( )** and **SUM ( )** check Chapter 7 for examples.

Let's list common number functions below.

Table 8.1: Common Number Functions

Oracle SQL	T-SQL	MySQL
CEIL ( )	CEILING ( )	CEIL ( ), CEILING ( )
FLOOR ( )	FLOOR ( )	FLOOR ( )
GREATEST ( )		GREATEST ( )
LEAST ( )		LEAST ( )
MOD ( )	%	MOD ( )
POWER (m, n)	POWER (m, n)	POW (m, n), POWER (m, n)
ROUND ( )	ROUND ( )	ROUND ( )
SQRT ( )	SQRT ( )	SQRT ( )
TRUNC ( )		TRUNC ( )

CEIL ( ) – Oracle and MySQL

CEILING ( ) – T-SQL

Description: Returns the smallest whole number greater than or equal to a specified number.

Syntax   **CEIL** (number)
                    CEILING (number)

Question 1: Write a query to find a whole number that is greater than or equal to 12.5.

Answers:

Oracle SQL	T-SQL	MySQL
SELECT **CEIL** (12.5) FROM dual; CEIL(12.5) 13	SELECT **CEILING** (12.5) Ceiling 13	SELECT **CEIL** (12.5); CEIL(12.5) 13

Figure 8.1 Query and output for question 1

FLOOR ( )

Description—Returns the largest whole number less than or equal to a specified number.

Syntax   **FLOOR** (number)

Question 2: Write a query to get a whole number that is less than or equal to 12.5.

Answers:

Oracle SQL	T-SQL	MySQL
SELECT **FLOOR** (12.5) FROM dual; FLOOR(12.5) 12	SELECT **FLOOR** (12.5) Floor; Floor 12	SELECT **FLOOR** (12.5); Floor (12.5) 12

Figure 8.2 Query and output for question 2

GREATEST ( ) – Oracle and MySQL

Description—Returns the greatest number or the largest character value in a list.

Syntax   **GREATEST** (a list of numbers or characters)

Question 3: Write a query to display the greatest number in a list of 4, 8 and 2.

Answers:

Oracle SQL	MySQL
SELECT **GREATEST** (4, 8, 2) FROM dual; GREATEST(4,8,2) 8	SELECT **GREATEST** (4, 8, 2); GREATEST(4,8,2) 8

Figure 8.3 Query and output for question 3

Question 4: Write a query to display the largest character value in a list of 'F', 'U', and 'B'.

Answers:

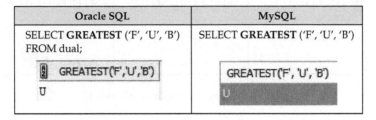

Oracle SQL	MySQL
SELECT **GREATEST** ('F', 'U', 'B') FROM dual;	SELECT **GREATEST** ('F', 'U', 'B')

Figure 8.4 Query and output for question 4

LEAST ( ) – Oracle and MySQL

Description: Returns the smallest number in a list.

Syntax   **LEAST** (a list of numbers)

Question 5: Write a query to display the smallest number in a list of 4, 8 and 2.

Answers:

Oracle SQL	MySQL
SELECT **LEAST** (4,8,2) FROM dual;	SELECT **LEAST** (4,8,2);
LEAST(4,8,2)	LEAST(4,8,2)
2	2

Figure 8.5 Query and output for question 5

MOD ( ) – Oracle and MySQL

          % Operator – T-SQL

Description: Returns the remainder of *Num1* divided by *Num2*.

Syntax   **MOD** (Num1, Num2)

                      Num1 % Num2 (T-SQL)

Question 6: Write a query to get the remainder of 10 modulo 4.

Answers:

Oracle SQL	T-SQL	MySQL
SELECT **MOD** (10, 4) FROM dual;	SELECT 10% 4 as Mod;	SELECT **MOD** (10, 4); SELECT 10 **MOD** 4;
MOD(10,4) 2	Mod 2	MOD (10, 4) 2

Figure 8.6 Query and output for question 6

POWER ( )

Description: Returns *Num1* raised to the *Num2*th power

Syntax　　　**POWER** (Num1, Num2)

Question 7: Write a query to display the 2 raised to 3 power.

Answers:

Oracle SQL	T-SQL	MySQL
SELECT **POWER** (2, 3) from dual;	SELECT **POWER** (2, 3) AS Power;	SELECT **POWER** (2, 3); SELECT **POW** (2, 3);
POWER(2,3) 8	Power 8	Power(2,3) 8

Figure 8.7 Query and output for question 7

ROUND ( ) Function

Description: Returns a number rounded to a certain digits after decimal points.

Syntax　　　**ROUND** (Number, d)
　　　　　　　Number – Column number or single number
　　　　　　　d – decimal places

Question 8: Write a query to round 267.389 with 2 digits after decimal points.

Answers:

Oracle SQL	T-SQL	MySQL
SELECT **ROUND** (267.389, 2) FROM dual; ROUND(267.389,2) 267.39	SELECT **ROUND** (267.389, 2) AS Round; Round 267.390	SELECT **ROUND** (267.389, 2) ROUND (267.389, 2) 267.39

Figure 8.8 Query and output for question 8

SQRT ( )

Description: Returns the square root of a number.

Syntax
SQRT (Number)

Question 9: Write a query to display a square root of 100.

Answers:

Oracle SQL	T-SQL	MySQL
SELECT **SQRT** (100) FROM dual; SQRT(100) 10	SELECT **SQRT** (100) AS Sqrt; Sqrt 10	SELECT **SQRT** (100) SQRT(100) 10

Figure 8.9 Query and output for question 9

TRUNC ( ) – Oracle & MySQL

Description: Returns *Number* truncated to *d* decimal places. The result number is not rounded.

Syntax   **TRUNC** (Number, d)
        d – decimal places

Question 10: Write a query to truncate 528.915 with 2 decimal places.

Answers:

Oracle SQL	MySQL
SELECT **TRUNC** (528.915, 2) FROM DUAL;	SELECT **TRUNCATE** (528.915, 2);
TRUNC(528.915,2) 528.91	TRUNCATE(528.915,2) 528.91

Figure 8.10 Query and output for question 10

COMMON STRING FUNCTIONS

Let's list common string functions below.

Table 8.2: Common String Functions

Oracle SQL	T-SQL	MySQL
CONCAT ( )	CONCAT ( )	CONCAT ( )
CONCAT ( ) using ||	CONCAT ( ) using +	
		FORMAT ( )
	LEFT ( )	LEFT ( )
INITCAP ( )		
LENGTH ( )	LEN ( )	LENGTH ( )
LOWER ( )	LOWER ( )	LOWER ( )
LPAD ( )		LPAD ( )
LTRIM ( )	LTRIM ( )	LTRIM ( )
REPLACE ( )	REPLACE ( )	REPLACE ( )
	RIGHT ( )	RIGHT ( )
RPAD ( )		RPAD ( )
RTRIM ( )	RTRIM ( )	RTRIM ( )
SUBSTR ( )	SUBSTRING ( )	SUBSTR ( ), SUBSTRING ( )
TRIM ( )		TRIM ( )
UPPER ( )	UPPER ( )	UPPER ( )

The Employees table and the Job table will be used for sample data in this section.

EMPLOYEE_ID	FIRST_NAME	LAST_NAME	EMAIL	PHONE_NUMBER	HIRE_DATE	JOB_ID	SALARY	MANAGER_ID	DEPARTMENT_ID
100	Douglas	Grant	DGRANT	650.507.9844	2008-01-23	SH_CLERK	2600.00	114	50
101	Adam	Fripp	AFRIPP	650.123.2234	2015-04-10	SH_MGR	8200.00	109	50
102	Jennifer	Whalen	JWHALEN	515.123.4444	2013-09-06	AD_ASST	4400.00	108	10
103	Michael	Hartstein	MHARTSTE	515.123.5555	2014-02-17	MK_MGR	13000.00	109	20
104	Pat	Fay	PFAY	603.123.6666	2005-08-01	MK_REP	6000.00	103	20
105	Susan	Mavris	SMAVRIS	515.123.7777	2012-06-22	HR_MGR	6500.00	109	40
106	Shelley	Higgins	SHIGGINS	515.123.8080	2012-05-26	SA_MGR	12008.00	109	80
107	William	Gietz	WGIETZ	515.123.8181	2012-02-20	SA_REP	8300.00	106	80
108	Steven	King	SKING	515.123.4567	2013-06-15	AD_PRES	24000.00	108	10
109	Lex	De Haan	LDEHAAN	515.123.4569	2011-01-23	AD_VP	17000.00	108	10
110	Bruce	Ernst	BERNST	590.423.4568	2017-02-21	IT_MGR	6000.00	109	60
111	Diana	Lorentz	DLORENTZ	590.423.5567	2017-01-18	IT_PROG	4200.00	110	60
112	Nancy	Greenberg	NGREENBE	515.124.4569	2013-03-22	FI_MGR	12008.00	109	90
113	Daniel	Faviet	DFAVIET	515.124.4169	2012-08-25	FI_CLERK	3000.00	112	90

Figure 8.11 The Employees table

JOB_ID	JOB_TITLE	MIN_SALARY	MAX_SALARY
AD_PRES	CEO	9000	25000
AD_VP	VICE President	8000	18000
AD_ASST	Admin Assistant	5000	6000
FI_CLERK	Finance Clerk	3000	4000
FI_MGR	Finance Manager	4000	5000
SA_REP	Sales Representative	3000	4000
SA_MGR	Sales Manager	4000	5000
SH_CLERK	Shipping Clerk	2500	4000
SH_MGR	Shipping Manager	4000	5000
IT_PROG	Programmer	4000	5500
IT_MGR	IT Manager	5000	6000
MK_CLERK	Marketing Clerk	3000	4000
MK_MGR	Marketing Manager	4000	5000
HR_MGR	Human Resource ...	4000	5000

Figure 8.12 The Job table

CONCAT ( )

Description: Concatenates two strings together.

Syntax   **CONCAT** (string1, string2)

Question 11: Write a query to concatenate first name and last name.

Answer (Oracle):

> SELECT **CONCAT** (First_Name, **CONCAT** ('', Last_Name)) AS FullName
> FROM Employees;

Answer (T-SQL & MySQL):

> SELECT **CONCAT** (First_Name, '', Last_Name) AS FullName
> FROM Employees;

Oracle SQL	T-SQL	MySQL
FULLNAME Douglas Grant Adam Fripp Jennifer Whalen	FullName Douglas Grant Adam Fripp Jennifer Whalen	FullName Douglas Grant Adam Fripp Jennifer Whalen

Figure 8.13 Query output for CONCAT function (top 3 rows)

CONCAT ( ) with Oracle || operator or T-SQL ' + ' operator

Syntax

> string1 || string2 (Oracle)
> string1 + string2 (T-SQL)

Question 12: Write a query to concatenate first name and last name using Oracle || operator or T-SQL "+" operators.

Answer (Oracle):

> SELECT First_Name ||'' || Last_Name AS FullName
> FROM Employees;

Answer (T-SQL):

> SELECT First_Name + '' + Last_Name AS FullName
> FROM Employees;

Oracle SQL	T-SQL
FULLNAME Douglas Grant Adam Fripp Jennifer Whalen	FullName Douglas Grant Adam Fripp Jennifer Whalen

Figure 8.14 Query output for question 12 (top 3 rows)

FORMAT ( ) – MySQL

Description: Rounds a numeric value to a number of decimal places. The result is a string.

Syntax　　　**FORMAT** (Number, d)

　　　　　　　d – decimal places

MySQL FORMAT function example:

　　　　SELECT **FORMAT**(623.7085, 2);

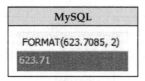

Figure 8.15 Query output for FORMAT example

LEFT ( ) – T-SQL & MySQL

Description: Gets a certain number of characters from the left.

Syntax　　　**LEFT** (string, length)

　　　　　　　length–length of specified number

Question 13: Get four characters from the left side of string "Database".

Answer (T-SQL):

　　　　SELECT **LEFT**('Database', 4) AS Left_Function;

Answer (MySQL):

　　　　SELECT **LEFT**('Database', 4);

T-SQL	MySQL
Left_Function Data	LEFT('Database', 4) Data

Figure 8.16 Query output for LEFT function

INITCAP ( ) – Oracle

Description: Changes the 1st character in each word to uppercase.

Syntax　　　**INITCAP** (string)

INITCAP function examples:

> SELECT **INITCAP**('oracle sql server mysql')
> FROM dual;

> SELECT **INITCAP**('ORACLE SQL SERVER MYSQL')
> FROM dual;

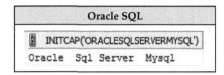

Figure 8.17 Query output for INITCAP function

LENGTH ( ) – Oracle & MySQL

LEN ( ) – T-SQL

Description: returns the length of the specified string.

Syntax	**LENGTH** (string)
	LEN (string)

Question 13: Write a query to display the country names that are greater than 10 characters in length.

Answers (Oracle & MySQL):

> SELECT COUNTRY_NAME
> FROM Country
> WHERE **LENGTH** (country_name) >= 10;

Answer (T-SQL):

> SELECT COUNTRY_NAME
> FROM Country
> WHERE **LEN** (country_name) >= 10;

Oracle SQL	T-SQL	MySQL
COUNTRY_NAME	COUNTRY_NAME	COUNTRY_NAME
Switzerland	Switzerland	Switzerland
Netherlands	Netherlands	Netherlands
United Kingdom	United Kingdom	United Kingdom
United States of America	United States of America	United States of America

Figure 8.18 Query output for LEN (LENGTH) function

LOWER ( )

Description: Converts all letters in the specified string to lowercase.

Syntax   **LOWER** (string)

Question 14: Write a query to display all the country names in lower case.

Answers (Oracle & MySQL):

>   SELECT **LOWER** (country_name)
>   FROM country;

Answers (T-SQL):

>   SELECT
>   **LOWER** (country_name) AS country_name
>   FROM country;

Oracle SQL	T-SQL	MySQL
LOWER(COUNTRY_NAME) argentina australia belgium	country_name argentina australia belgium	LOWER (country_name) argentina australia belgium

Figure 8.19 Query output for LOWER function (top 3 rows)

LPAD ( ) – Oracle and MySQL

Description: Adds a set of characters to the left side of a string

Syntax   **LPAD** (string_1, padded_length, pad_characters)

Question 14: Add the area code '706' to '352-7100'.

Answer (Oracle):

>   SELECT **LPAD** ('352-7100', 12, '706-')
>   FROM dual;

Answer (MySQL):

> SELECT **LPAD** ('352-7100', 12, '706-');

Oracle SQL	MySQL
▦ LPAD('352-7100',12,'706-') 706-352-7100	LPAD ('352-7100', 12, '706-') 706-352-7100

Figure 8.20 Query output for LPAD function

LTRIM ( )

Description: Removes a set of characters from the left side of a string.

Syntax **LTRIM** (string_1, trim_characters) (Oracle)
LTRIM (string_1) (T-SQL & MySQL) – Removes space characters from the left side of a string.

Example (Oracle): Remove the area code 706 from '706-352-7100'.

> SELECT **LTRIM** ('706-352-7100', '706-')
> FROM dual;

Example (T-SQL): Remove the left spaces from '706-352-7100'.

> SELECT **LTRIM** ('706-352-7100') AS LTRIM_Function;

Example (MySQL): Remove the left spaces from '706-352-7100'.

> SELECT **LTRIM** ('706-352-7100');

Oracle SQL	T-SQL	MySQL
▦ LTRIM('706-352-7100','706-') 352-7100	LTRIM_Function 706-352-7100	LTRIM (' 706-352-7100') 706-352-7100

Figure 8.21 Query output for LTRIM function

REPLACE ( )

Description: Replaces part of a string with specified character(s).

Syntax **REPLACE** ('string1', 'str_to_be_seached', 'str_to_to_replaced')

Question 15: Write a query to replace '–' with '.' for the phone field.

Answer:

SELECT first_name, last_name, **REPLACE** (phone, '–', '.') as Phone
FROM employees;

Oracle SQL			T-SQL			MySQL		
FIRST_NAME	LAST_NAME	PHONE	first_name	last_name	Phone	first_name	last_name	Phone
Douglas	Grant	650.507.9844	Douglas	Grant	650.507.9844	Douglas	Grant	650.507.9844
Adam	Fripp	650.123.2234	Adam	Fripp	650.123.2234	Adam	Fripp	650.123.2234
Jennifer	Whalen	515.123.4444	Jennifer	Whalen	515.123.4444	Jennifer	Whalen	515.123.4444

Figure 8.22 Query output for REPLACE function (Top 3 rows)

RIGHT ( ) – T-SQL & MySQL

Desciption: Get a certain number of characters from the right.

Syntax   **RIGHT** (string, length)
                 length – length of specified number

Question 16: Get four characters from right side of string "Database".

Answer (T-SQL):

SELECT **RIGHT**('Database', 4) AS Right_Function;

Answer (MySQL):

SELECT **RIGHT**('Database', 4);

T-SQL	MySQL
Right_Function	RIGHT('Database', 4)
base	base

Figure 8.23 Query output for RIGHT function

RPAD ( ) – Oracle and MySQL

Description—Adds a set of characters to the right side of a string.

Syntax   **RPAD** (string_1, padded_length, pad_characters)

Question 17: Add 'vision' to the right side of string 'Tele'.

Answer (Oracle):

SELECT **RPAD** ('Tele', 10, 'vision') FROM dual;

Answer (MySQL):

> SELECT **RPAD** ('Tele', 10, 'vision');

Oracle SQL	MySQL
▤ RPAD('TELE', 10, 'VISION') Television	RPAD ('Tele', 10, 'vision') Television

Figure 8.24 Query output for RPAD function

RTRIM ( )

Description: Removes a set of characters from the right side of a string.

Syntax (Oracle)

> **RTRIM** (string_1, trim_characters)

Syntax (T-SQL & MySQL)

> **RTRIM** (string_1) – Removes space characters from the right side of a string.

Example (Oracle): Remove 0s in '57800' .

> SELECT **RTRIM** ('57800', '0')
> FROM dual;

Example (T-SQL & MySQL): Remove right spaces from 'Television'.

> SELECT **RTRIM** ('Television') AS 'RTRIM';

Oracle SQL	T-SQL	MySQL
▤ RTRIM('57800','0') 578	RTRIM Television	RTRIM Television

Figure 8.25 Query output for RTRIM function

SUBSTR ( )

SUBSTRING ( )

Description: Extract a substring from a start position with length in a string.

Syntax   **SUBSTR** (string, position, length) (Oracle, MySQL)
SUBSTRING (string, position, length) (T-SQL, MySQL)

> position – integer
> length – integer

Question 18: Write a query to display the first three characters for the last name field.

Answers (Oracle & MySQL):

SELECT Last_Name, **SUBSTR** (Last_Name, 1, 3)
FROM employees;

Answers (T-SQL & MySQL):

SELECT Last_Name, **SUBSTRING** (Last_Name, 1, 3)
FROM employees;

Oracle SQL		T-SQL		MySQL	
LAST_NAME	SUBSTR(LAST_NAME,1,3)	Last_Name	(No column name)	Last_Name	SUBSTRING(Last_Name, 1, 3)
Grant	Gra	Grant	Gra	Grant	Gra
Fripp	Fri	Fripp	Fri	Fripp	Fri
Whalen	Wha	Whalen	Wha	Whalen	Wha

Figure 8.26 Query output for SUBSTR (SUBSTRING) function (Top 3 rows)

UPPER ( )

Description: Converts all letters in the specified string to upper case.

Syntax   **UPPER** (string)

Question 17: Write a query to display all the country names in upper case.

Answers (Oracle & MySQL)

SELECT **UPPER** (country_name)
FROM country;

Answers (T-SQL)

SELECT **UPPER** (country_name) AS country_name
FROM country;

Oracle SQL	T-SQL	MySQL
UPPER(COUNTRY_NAME)	country_name	UPPER (country_name)
ARGENTINA	ARGENTINA	ARGENTINA
AUSTRALIA	AUSTRALIA	AUSTRALIA
BELGIUM	BELGIUM	BELGIUM

Figure 8.27 Query output for UPPER function (Top 3 rows)

Common Date and Time Functions

Let's list common date and time functions below.

Table 8.3 Date and Time Functions

Oracle SQL	T-SQL	MySQL
CURRENT_TIMESTAMP	CURRENT_TIMESTAMP	CURRENT_TIMESTAMP
Add_Months ( )	DATEADD ( )	DATE_ADD ( )
EXTRACT ( )	DATEPART ( )	EXTRACT ( )
CURRENT_DATE	GETDATE ( )	CURRENT_DATE
MONTHS_BETWEEN ( )	DATEDIFF ( )	PERIOD_DIFF ( )
SYSDATE	SYSDATETIME ( )	SYSDATE ( )

CURRENT_TIMESTAMP

Example (Oracle):

SELECT **CURRENT_TIMESTAMP** FROM dual;

Example (T-SQL):

SELECT **CURRENT_TIMESTAMP** AS 'Current_Time';

Example (MySQL):

SELECT **CURRENT_TIMESTAMP;**

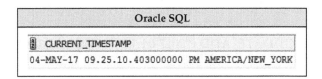

Figure 8.28 Query output for Oracle CURRENT_TIMESTAMP

T-SQL	MySQL
Current_Time	CURRENT_TIMESTAMP
2017-05-04 21:20:14.643	2017-05-04 21:16:13

Figure 8.29 Query output for T-SQL and MySQL CURRENT_TIMESTAMP

Add_Months ( ) – Oracle

Description: Returns a date with n months after the specified date.

Syntax
Add_Month (date, n)

DATEADD ( ) – T-SQL

Syntax
DATEADD (month, n, date)

DATE_ADD ( ) – MySQL

Syntax
DATE_ADD (date , INTERVAL n MONTH)

Question 18: Write a query to add one month to April 3rd, 2017

Answer (Oracle):

> SELECT **ADD_MONTHS** ('03-Apr-17', 1)
> FROM DUAL;

Answer (T-SQL):

> SELECT **DATEADD** (month, 1, '2017/04/03') AS Date;

Answers (MySQL):

> SELECT **DATE_ADD** ("2017-04-17", INTERVAL 1 MONTH)

Oracle SQL	T-SQL	MySQL
ADD_MONTHS('03-APR-17,1) 03-MAY-17	Date 2017-05-03 00:00:00.000	DATE_ADD("2017-04-03", INTERVAL 1 MONTH) 2017-05-03

Figure 8.30 Query output for Add month function

EXTRACT ( ) – Oracle & MySQL

Description: Extracts the value of a specified date time field from a date time field.

Syntax   **EXTRACT** (YEAR/MONTH/WEEK/DAY/HOUR/MINUTE
            FROM DATE)

DATEPART ( ) – T-SQL

Question 19: Extract YEAR from April 4th, 2017.

Answer (Oracle):

> SELECT **EXTRACT** (YEAR FROM DATE '2017-04-03') "Year"
> FROM DUAL;

Answer (T-SQL):

> SELECT **DATEPART** (year, '2017/04/03') AS Year;

Answer (MySQL):

> SELECT **EXTRACT** (YEAR FROM DATE '2017-04-03') As Year;

Oracle SQL	T-SQL	MySQL
Year 2017	Year 2017	Year 2017

Figure 8.31 Query output for EXTRACT function

CURRENT_DATE – Oracle

GETDATE ( ) – T-SQL
CURRENT_DATE ( ) – MySQL

Question 20: Write a query to display the current date.

Answer (Oracle):

> SELECT **CURRENT_DATE**
> FROM dual;

Answer (T-SQL):

> SELECT **GETDATE( )** AS Date;

Answer (MySQL):

> SELECT **CURRENT_DATE( );**

Oracle SQL	T-SQL	MySQL
◊ CURRENT_DATE 28-OCT-16	Date 2017-05-04 21:31:31.713	CURRENT_DATE( ) 2017-05-04

Figure 8.32 Query output for CURRENT_DATE function

MONTHS_BETWEEN ( ) – Oracle

Style   **MONTHS_BETWEEN** (date1, date2)

DATEDIFF ( ) – T-SQL

Style   **DATEDIFF** (month, date1, date2)

PERIOD_DIFF ( ) – MySQL

Style   **PERIOD_DIFF**(date1, date2)

Question 21: Write a query to display the number of months between 12/31/2015 to 12/02/2016.

Answer (Oracle):

> SELECT **MONTHS_BETWEEN**
> (TO_DATE ('12-02-2016','MM-DD-YYYY'), TO_DATE ('12-31-2015','MM-DD-YYYY')) AS Month
> FROM dual ;

Answer (T-SQL):

> SELECT **DATEDIFF** (month, '12-31-2015', '12-02-2016') as Month

Answer (MySQL):

> SELECT **PERIOD_DIFF** ('201612', '201512') as Months

Oracle SQL	T-SQL	MySQL
MONTH 11.064516:	Month 12	Months 12

Figure 8.33 Query output for MONTH_BETWEEN function

SYSDATE – Oracle

SYSDATETIME ( ) – T-SQL

SYSDATE ( ) – MySQL

Question 22: Write a query to display the system date.

Answer (Oracle):

> SELECT sysdate
> FROM dual;

Answer (T-SQL):

> SELECT sysdatetime( );

Answer (MySQL):

SELECT sysdate ( );

Oracle SQL	T-SQL	MySQL
ⒶⒷ SYSDATE 04-MAY-17	sysdatetime 2017-05-04 21:42:40.3616068	sysdate() 2017-05-04 21:44:25

Figure 8.34 Query output for SYSDATE function

Conversion Functions

Table 8.4 Conversion Functions

Oracle SQL	T-SQL	MySQL
CAST ( )	CAST ( )	CAST ( )
TO_DATE ( )	CONVERT ( )	STR_TO_DATE ( )

CAST ( )

Description: converts an expression from one datatype to another datatype.

Syntax   **CAST** (expression AS data_type)

Question 23: Write a query to change 356.78 to an integer number.

Answer (Oracle):

SELECT **CAST** (356.78 as int)
FROM dual;

Answer (T-SQL):

SELECT **CAST** (356.78 AS int) AS CAST;

Answer (MySQL):

SELECT CAST(356.78 SIGNED INTEGER);

Oracle SQL	T-SQL	MySQL
CAST(356.78ASINT) 357	CAST 356	CAST(356.78 as SIGNED INT) 357

Figure 8.35 Query output for CAST function

TO_DATE ( ) – Oracle

Style   **TO_DATE** (*string, format*)

Oracle To_Date function format is listed in the table below.

Table 8.5 Oracle To-Date format

Format	Description
YYYY	4-digit year
YY	2-digit year
MON	January–December
MM	1–12
DY	Sun–Sat
DD	0–23
HH24	1–31
HH or HH12	1–12
MI	0–59
SS	0–59

Example 1:   SELECT **TO_DATE**('2016/10/25', 'YYYY/MM/DD')
FROM dual;

```
TO_DATE('2016/10/25','YYYY/MM/DD')
25-OCT-16
```

Figure 8.36 Query output for Oracle To_Date function

Example 2:   SELECT **TO_DATE**('20161026', 'YYYYMMDD')
FROM dual;

```
TO_DATE('20161026','YYYYMMDD')
26-OCT-16
```

Figure 8.37 Query output for Oracle To_Date function example 2

CONVERT ( ) – T-SQL

Style   **CONVERT** *(data_type, expression, style)*

T-SQL date conversion styles and samples are listed below:

Table 8.6 T-SQL CONCERT( ) function styles

Style Number	Sample
101	12/16/2016
102	2016.12.16
103	16/12/2016
104	16.12.2016
105	16-12-2016
106	16 Dec 2016
DATE	YYYY-MM-DD
DATETIME	YYYY-MM-DD HH:MI:SS

T-SQL Date Conversion Examples:

Example 1: SELECT **CONVERT** (varchar, getdate ( )) AS Date;

Date

May 4 2017 9:46PM

Figure 8.38 Query output for CONERT example 1

Example 2: SELECT **CONVERT** (varchar, getdate ( ), 101) AS Date;

Date

05/04/2017

Figure 8.39 Query output for CONERT example 2

Example 3: SELECT **CONVERT** (varchar, getdate ( ), 106) AS Date;

Date

04 May 2017

Figure 8.40 Query output for CONERT example 3

STR_TO_DATE ( ) – MySQL

Style   **STR_TO_DATE** (*String, Format*);

MySQL date conversion formats are listed below:

Format	Description
%Y	4-digit year
%y	2-digit year
%b	Abbreviated month (Jan–Dec)
%M	Month name (January–December)
%m	Month (0–12)
%a	Abbreviated day (Sun–Sat)
%d	Day (0–31)
%H	Hour (0–23)
%h	Hour (01–12)
%i	Minutes (0–59)
%s	Seconds (0–59)

Example 1: SELECT **STR_TO_DATE**('May 01 2017', '%M %d %Y') AS date;

Figure 8.41 Query output for STR_TO_DATE function example 1

Example 2: SELECT STR_TO_DATE('2016,5,20 04,20,35', '%Y, %m, %d %h,%i, %s') AS date;

Figure 8.42 Query output for STR_TO_DATE function example 2

Summary

Chapter 8 covers the following:

- Understanding common number functions.
- Using common character functions
- How to use common date and time functions
- Using conversion functions

Exercises

8.1

Write a query to display the year portion of the system date.

8.2

Write a query to display rounded 682.3547 to two digits after decimal points.

8.3

Write a query to display the 8th through 10th characters of the string "Oracle SQL Developer".

Chapter 9

Advanced SQL

In this chapter, you will learn how to use the following SQL commands:

1. **UNION, UNION ALL**
2. **INTERSECT** (Oracle and T-SQL), **IN** (MySQL)
3. **EXCEPT** (T-SQL), **MINUS** (Oracle), **NOT IN** (MySQL)
4. **ROWNUM** (Oracle), **TOP** (T-SQL) **and LIMIT** (MySQL)
5. **Subquery**
6. **CASE**
7. **SEQUENCE** (Oracle), **IDENTITY** (T-SQL), **AUTO_INCREMENT** (MySQL)

We will use Customers and Locations tables for sample data here.

ID	First Name	Last Name	City	Country	Phone
1	Howard	Bell	Seattle	USA	(678)-555-7629
2	Alice	Carter	Southlake	USA	(617)-213-6874
3	Carine	Schmitt	Nantes	France	40.32.21.21
4	Paolo	Accorti	Torino	Italy	011-4988260
5	Helen	Bennett	Barcelona	Spain	(93)-203-4560

Figure 9.1 Customers table

LOCATION_ID	STREET_ADDRESS	POSTAL_CODE	CITY	STATE_PROVINCE	COUNTRY_ID
1300	9450 Kamiya-cho	6823	Hiroshima		JP
1400	2014 Jabberwocky Rd	26192	Southlake	Texas	US
1500	2011 Interiors Blvd	99236	South San Francisco	California	US
1600	2007 Zagora St	50090	South Brunswick	New Jersey	US
1700	2004 Charade Rd	98199	Seattle	Washington	US
1800	147 Spadina Ave	M5V 2L7	Toronto	Ontario	CA
1900	6092 Boxwood St	YSW 9T2	Whitehorse	Yukon	CA
2000	40-5-12 Laogianggen	190518	Beijing		CN
2200	12-98 Victoria Street	2901	Sydney	New South Wales	AU
2400	8204 Arthur St		London		UK

Figure 9.2 Locations table

1. UNION, UNION All

Description

UNION:	Returns a distinct list of rows from two tables.
UNION ALL:	Returns all rows from both tables.

Syntax

SELECT column(s) FROM table1

UNION
SELECT column(s) FROM table2
SELECT column(s) FROM table1

UNION ALL
SELECT column(s) FROM table2

Note: Each SELECT statement within the UNION must have the same number of columns. The columns must also have similar data types. The columns in each SELECT statement must be in the same order.

Question 1: Write a query to combine the distinct cities in Customers and Locations tables.

Answer:

SELECT City
FROM Customer
UNION
SELECT City
FROM Locations
ORDER BY City;

Oracle SQL	T-SQL	MySQL
CITY	City	City
Barcelona	Barcelona	Barcelona
Beijing	Beijing	Beijing
Hiroshima	Hiroshima	Hiroshima
London	London	London
Nantes	Nantes	Nantes
Seattle	Seattle	Seattle
South Brunswick	South Brunswick	South Brunswick
South San Francisco	South San Francisco	South San Francisco
Southlake	Southlake	Southlake
Sydney	Sydney	Sydney
Torino	Torino	Torino
Toronto	Toronto	Toronto
Whitehorse	Whitehorse	Whitehorse

Figure 9.3 Query output for question 1

Question 2: Write a query to combine the cities in Customer and Location table.

Answers:

> SELECT City
> FROM Customer
> **UNION ALL**
> SELECT City
> FROM Locations
> ORDER BY City;

Oracle SQL	T-SQL	MySQL
CITY	City	City
Barcelona	Barcelona	Barcelona
Beijing	Beijing	Beijing
Bombay	Hiroshima	Hiroshima
Hiroshima	London	London
London	Nantes	Nantes
Nantes	Seattle	Seattle
Seattle	Seattle	Seattle
Seattle	South Brunswick	South Brunswick
South Brunswick	South San Francisco	South San Francisco
South San Francisco	Southlake	Southlake
Southlake	Southlake	Southlake
Southlake	Sydney	Sydney
Sydney	Torino	Torino
Torino	Toronto	Toronto
Toronto	Whitehorse	Whitehorse
Whitehorse		

Figure 9.4 Query output for question 2

2. INTERSECT (Oracle and T-SQL), IN (MySQL)

Description: Returns only rows that exist in both tables

Syntax

> SELECT column(s) FROM table1
> **INTERSECT**
> SELECT column(s) FROM tables

Use an intersect operator to returns rows that are common between two tables; it returns unique rows that exist in both the first and second query. This operation is useful when you want to find results that are common between two queries. **INTERSECT** has an equivalent MySQL statement **IN,** which can also be used in Oracle and T-SQL.

Question 3: Write a query to find cities that exist in both Customer and Locations tables.

Answers (Oracle & T-SQL):

> SELECT City
> FROM Customer
> **INTERSECT**
> SELECT City
> FROM Locations;

Answer (MySQL):

> SELECT distinct City
> FROM customer
> WHERE (city) **IN**
>   (SELECT City
>   FROM Locations)

Oracle SQL	T-SQL	MySQL
CITY	City	City
Seattle	Seattle	Seattle
Southlake	Southlake	Southlake

Figure 9.5 Query output for question 3

3. MINUS (Oracle), EXCEPT (T-SQL), NOT IN (MySQL)

Description: Returns all rows in the first SELECT statement but excludes those by the second SELECT statement.

Syntax

> SELECT col1, col2, …
> FROM table1
> **MINUS or EXCEPT**
> SELECT col1, col2, …
> FROM table2

Like INTERSECTION, EXCEPT (MINUS) has an equivalent MySQL statement **NOT IN,** which can also be used in Oracle and T-SQL.

Question 4: Write a query to find cities that exist in Locations table but not in Customer table.

Answers: See Table 9.1.

Table 9.1 Answers for the question

Oracle SQL	T-SQL	MySQL
SELECT City FROM Locations **MINUS** SELECT City FROM Customer;	SELECT City FROM Locations **EXCEPT** SELECT City FROM Customer;	SELECT distinct City FROM Locations WHERE (city) **NOT IN** (SELECT City FROM Customer)

Oracle SQL	T-SQL	MySQL
CITY Beijing Hiroshima London South Brunswick South San Francisco Sydney Toronto Whitehorse	City Beijing Hiroshima London South Brunswick South San Francisco Sydney Toronto Whitehorse	City Beijing Hiroshima London South Brunswick South San Francisco Sydney Toronto Whitehorse

Figure 9.6 Query output for question 4

4. ROWNUM (Oracle), TOP (T-SQL) and LIMIT (MySQL)

Description: Specifies the number of records to return

Oracle Style   SELECT column_name(s)
FROM table_name
WHERE **ROWNUM** <= number;

T-SQL Style   SELECT **TOP** number I percent column_name(s)
FROM table_name;

MySQL Style   SELECT column_name(s)
FROM table_name
LIMIT number;

ROWNUM is a special virtual column in an Oracle Database that gets many people into trouble. When you learn what it is and how it works, however, it can be very useful.

ROWNUM is available in a query, but is not part of the table. ROWNUM will be assigned the numbers 1, 2, 3, 4, ... N, where N is the number of rows record set. ROWNUM can be used as part of the where clause of the query to return specific rows. ROWNUM value is not assigned permanently to a row (this is a common misconception). Queries that use < (less than) or > (greater than) on ROWNUM will not always work; you must use <= (less than or equal to) or >= (greater than or equal to).

For example,

SELECT **ROWNUM**, firstname
FROM customer;

ROWNUM	FIRSTNAME
1	Jason
2	Doug
3	Maria

Figure 9.7 Query output for ROWNUM example

Questions 5: Display the first ten rows from the Country table.

Answers (Oracle):

SELECT *
FROM Country
WHERE **ROWNUM** <= 10;

COUNTRY_ID	COUNTRY_NAME	REGION_ID
AR	Argentina	2
AU	Australia	3
BE	Belgium	1
BR	Brazil	2
CA	Canada	2
CH	Switzerland	1
CN	China	3
DE	Germany	1
DK	Denmark	1
EG	Egypt	4

Figure 9.8 Query output for question 5 (Oracle)

Answers (T-SQL):

> SELECT **TOP** 10 *
> FROM Country;

COUNTRY_ID	COUNTRY_NAME	REGION_ID
AR	Argentina	2
AU	Australia	3
BE	Belgium	1
BR	Brazil	2
CA	Canada	2
CH	Switzerland	1
CN	China	3
DE	Germany	1
DK	Denmark	1
EG	Egypt	4

Figure 9.9 Query output for question 5 (T-SQL)

Answers (MySQL):

> SELECT *
> FROM Country **LIMIT** 10;

COUNTRY_ID	COUNTRY_NAME	REGION_ID
AR	Argentina	2
AU	Australia	3
BE	Belgium	1
BR	Brazil	2
CA	Canada	2
CH	Switzerland	1
CN	China	3
DE	Germany	1
DK	Denmark	1
EG	Egypt	4

Figure 9.10 Query output for question 5 (MySQL)

5. Subquery

Description: A Subquery is a SQL query nested inside a larger query. Subqueries should be placed within parenthesis. Subqueries can appear in the SELECT, FROM or WHERE clauses of the main query and create temporary virtual tables usable by the main query.

Style   SELECT column(s)
             FROM table1
             WHERE value IN
             (SELECT column-name
             FROM table2
             WHERE condition)

Question 6: Write a query to find the employees whose salary is greater than the average salary.

Answer: SELECT first_name, last_name, dept_ID, salary
             FROM employees
             WHERE salary >
             (SELECT AVG(salary)
             FROM employees);

Oracle output

first_name	last_name	dept_ID	salary
Michael	Hartstein	20	13000.00
Shelley	Higgins	80	12008.00
Steven	King	10	24000.00
Lex	De Haan	10	17000.00
Nancy	Greenberg	90	12008.00

T-SQL output

FIRST_NAME	LAST_NAME	DEPT_ID	SALARY
Michael	Hartstein	20	13000
Shelley	Higgins	80	12008
Steven	King	10	24000
Lex	De Haan	10	17000
Nancy	Greenberg	90	12008

MySQL output

first_name	last_name	dept_ID	salary
Michael	Hartstein	20	13000.00
Shelley	Higgins	80	12008.00
Steven	King	10	24000.00
Lex	De Haan	10	17000.00
Nancy	Greenberg	90	12008.00

Figure 9.11 Query output for questions 6

Question 7: Write a query to find the employees who works in the Sales department.

Answer:

SELECT Employee_ID, First_Name, Last_Name, Dept_ID
FROM employees
WHERE dept_id IN
(SELECT dept_id
FROM departments
WHERE dept_name='Sales');

Oracle output

EMPLOYEE_ID	FIRST_NAME	LAST_NAME	DEPT_ID
107	William	Gietz	80
106	Shelley	Higgins	80

T-SQL output

Employee_ID	First_Name	Last_Name	Dept_ID
106	Shelley	Higgins	80
107	William	Gietz	80

MySQL output

Employee_ID	First_Name	Last_Name	Dept_ID
106	Shelley	Higgins	80
107	William	Gietz	80

Figure 9.12 Query output for question 7

6. CASE

Description: the CASE statement has the functionality of an IF-THEN-ELSE statement.

Syntax 1

CASE
WHEN condition_1 THEN result_1
WHEN condition_2 THEN result_2 ...
WHEN condition_n THEN result_n
ELSE result
END

Syntax 2

> **CASE** expression
> **WHEN** value_1 THEN result_1
> **WHEN** value_2 THEN result_2 ...
> **WHEN** value_n THEN result_n
> **ELSE** result
> **END**

The CASE statement always goes in the SELECT clause. CASE must include the following components: WHEN, THEN, and END. ELSE is an optional component.

You can make any conditional statement using any conditional operator (like WHERE) between WHEN and THEN. This includes stringing together multiple conditional statements using AND and OR.

You can include multiple WHEN statements, as well as an ELSE statement to deal with any unaddressed conditions.

CASE Syntax 1 Example:

> SELECT dept_name,
> **CASE**
>   **WHEN** location_id = 1400
>   OR location_id = 1500
>   OR location_id = 1700
>   OR location_id = 2500
>   OR location_id = 2700 THEN 'USA'
>   **WHEN** location_id = 1800 THEN 'Canada'
>   **WHEN** location_id = 2400 THEN 'UK'
> **END** "location"
> FROM Departments;

CASE Style 2 Example:

> SELECT dept_name,
> **CASE** location_id
>   **WHEN** 1400 THEN 'USA'
>   **WHEN** 1500 THEN 'USA'
>   **WHEN** 1700 THEN 'USA'
>   **WHEN** 2500 THEN 'USA'
>   **WHEN** 2700 THEN 'USA'
>   **WHEN** 1800 THEN 'Canada'
>   **WHEN** 2400 THEN 'UK'
> **END** "LOCATION"
> FROM Departments;

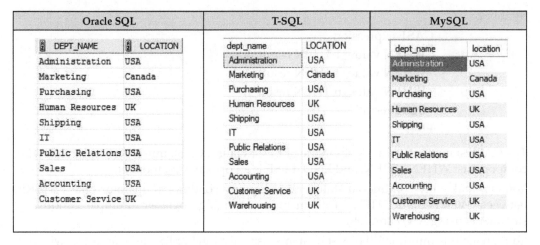

Figure 9.13 Query output for CASE Examples

7. SEQUENCE (Oracle), IDENTITY (T-SQL), AUTO_INCREMENT (MySQL)

Description: We would like the value of the primary key field to be created automatically every time a new record is inserted.

Oracle Style

> CREATE **SEQUENCE** sequence_name
> [**START WITH** start_num]
> [**INCREMENT BY** increment_num]

T-SQL Style

> **IDENTITY** (seed, increment)
> > seed - the initial number
> > increment - the interval

MySQL Style

> **AUTO_INCREMENT**: By default, the beginning value is 1, and it will increment by 1 for each new record.

Oracle SEQUENCE Example

Step 1—Create a SEQUENCE called seq_customer that starts from 100. Every time we insert a record the seq_customer.NEXTVAL generates a new value starts from 100:

> CREATE **SEQUENCE** seq_customer START WITH 100;

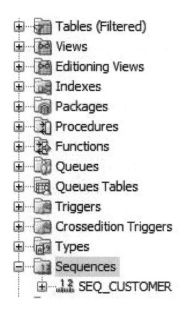

Figure 9.14 CREATE SEQUENCE in Oracle

Step 2—Insert a customer "Jason Brown" to the customer table. The seq_customer. NEXTVAL generates a new value starts from 100:

> INSERT INTO customer (id, FirstName, LastName)
> VALUES (seq_customer.**NEXTVAL**, 'Jason','Brown');

ID	FIRSTNAME	LASTNAME
1	Howard	Bell
2	Alice	Carter
3	Carine	Schmitt
4	Paolo	Accorti
5	Helen	Bennett
100	Jason	Brown

Figure 9.15 A customer record is inserted

T-SQL INDENTITY Example

Step 1—Create a table called STATE and set the initial number 1 and the interval value 1.

```
Create Table STATE
(
    ID int NOT NULL IDENTITY(1, 1),
    StateName varchar(30)
)
```

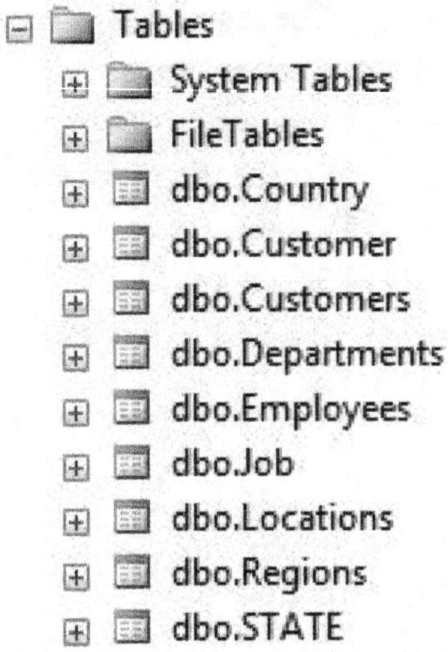

```
☐ 🗀 Tables
    ⊞ 🗀 System Tables
    ⊞ 🗀 FileTables
    ⊞ 🎛 dbo.Country
    ⊞ 🎛 dbo.Customer
    ⊞ 🎛 dbo.Customers
    ⊞ 🎛 dbo.Departments
    ⊞ 🎛 dbo.Employees
    ⊞ 🎛 dbo.Job
    ⊞ 🎛 dbo.Locations
    ⊞ 🎛 dbo.Regions
    ⊞ 🎛 dbo.STATE
```

Figure 9.16 The STATE table (SQL Server)

Step 2—Insert two states to the table. It is not necessary to insert IDs as INDENTIY will automatically create IDs one by one.

```
INSERT INTO STATE (StateName) VALUES ('Utah');
INSERT INTO STATE (StateName) VALUES ('Maryland');
```

Step 3—Check the result.

```
SELECT * FROM STATE
```

ID	StateName
1	Utah
2	Maryland

Figure 9.17 Step 3 query output

MySQL AUTO_INCREMENT Example

Step 1—Create a table called STATE with a primary key AUTO_INCREMENT. The initial number and the interval value are 1 by default.

> Create Table STATE
> (
>   ID int NOT NULL Primary Key **AUTO_INCREMENT**,
>   StateName varchar(30)
>)

Figure 9.18 The STATE table (MySQL)

To set the AUTO_INCREMENT with another starting value, use the following SQL statement:

> ALTER TABLE STATE AUTO_INCREMENT=50;

Step 2—Insert two states to the table. It is not necessary to insert IDs as AUTO_INCREMENT will create IDs.

> INSERT INTO STATE (StateName) VALUES ('Utah');
> INSERT INTO STATE (StateName) VALUES ('Maryland');

Step 3—Check the result.

> SELECT * FROM STATE;

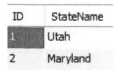

Figure 9.19 Step 3 query output (MySQL)

Summary

Chapter 9 covers the following:

- Manipulating Union, Union All commands
- Using INTERSECT (Oracle and T-SQL) and IN (MySQL) commands
- How to use Except (T-SQL), MINUS (Oracle) and NOT IN (MySQL) commands
- Understanding ROWNUM (Oracle), TOP (T-SQL) and LIMIT (MySQL) commands
- Defining Subquery
- Understanding Case command
- Using Sequence (Oracle), Identity (T-SQL) and Auto_Increment (MySQL) to generate sequence numbers.

Exercises

9.1

Write a query to combine the names in the customers table with the names in the employees table.

9.2

Modify the query in 9-1 to sort the results by last name.

Chapter 10

Joins

In relational database design Employee and Department are two entities. Employee related data is saved to the Employees table while department related data is saved to the Departments table. For linking purpose the Department_ID is created between the two tables.

If we want to display employee names and department names it is not possible to only use the Employees table or the Departments table. To list the department names after the employee names we need to use SQL JOIN.

There are four basic SQL joins: (Inner) Join, Left Join, Right Join, and Full Join. The most useful join is (Inner) Join.

Table 10.1 Common Join types

(INNER) JOIN	Get records that have matching values in both tables.
LEFT JOIN	Get all records from the table1 (**LEFT** table1) and the matched records from the table2 (**RIGHT** table2). If no match the result is NULL from the table2.
RIGHT JOIN	Get all records from the table2 (**RIGHT** table1) and the matched records from the table1 (**LEFT** table2). If no match the result is NULL from the table1.
FULL JOIN	Get all the rows from both the table1 and table2.

Syntax

SELECT table1.col_name, table2.col_name
FROM table1
(INNER) JOIN table2 **ON** table1.col_name = table2.col_name;

SELECT table1.col_name, table2.col_name
FROM table1
LEFT JOIN table2 **ON** table1.col_name = table2.col_name;

SELECT table1.col_name, table2.col_name
FROM table1
RIGHT JOIN table2 **ON** table1.col_name = table2.col_name;

SELECT table1.col_name, table2.col_name
FROM table1
FULL JOIN table2 **ON** table1.col_name = table2.col_name;

Older JOIN Syntax

SELECT table1.col_name, table2.col_name
FROM table1, table2
WHERE table1.col_name = table2.col_name;

JOIN with USING clause (Oracle)

Syntax   SELECT table1.col_name, table2.col_name
FROM table1
JOIN table2 **USING** (join_col_name);

Let's use the Employees, the Departments and the Locations tables for JOIN examples.

EMPLOYEE_ID	FIRST_NAME	LAST_NAME	EMAIL	PHONE_NUMBER	HIRE_DATE	JOB_ID	SALARY	MANAGER_ID	DEPARTMENT_ID
100	Douglas	Grant	DGRANT	650.507.9844	2008-01-23	SH_CLERK	2600.00	114	50
101	Adam	Fripp	AFRIPP	650.123.2234	2015-04-10	SH_MGR	8200.00	109	50
102	Jennifer	Whalen	JWHALEN	515.123.4444	2013-09-06	AD_ASST	4400.00	108	10
103	Michael	Hartstein	MHARTSTE	515.123.5555	2014-02-17	MK_MGR	13000.00	109	20
104	Pat	Fay	PFAY	603.123.6666	2005-08-01	MK_REP	6000.00	103	20
105	Susan	Mavris	SMAVRIS	515.123.7777	2012-06-22	HR_MGR	6500.00	109	40
106	Shelley	Higgins	SHIGGINS	515.123.8080	2012-05-26	SA_MGR	12008.00	109	80
107	William	Gietz	WGIETZ	515.123.8181	2012-02-20	SA_REP	8300.00	106	80
108	Steven	King	SKING	515.123.4567	2013-06-15	AD_PRES	24000.00	108	10
109	Lex	De Haan	LDEHAAN	515.123.4569	2011-01-23	AD_VP	17000.00	108	10
110	Bruce	Ernst	BERNST	590.423.4568	2017-02-21	IT_MGR	6000.00	109	60
111	Diana	Lorentz	DLORENTZ	590.423.5567	2017-01-18	IT_PROG	4200.00	110	60
112	Nancy	Greenberg	NGREENBE	515.124.4569	2013-03-22	FI_MGR	12008.00	109	90
113	Daniel	Faviet	DFAVIET	515.124.4169	2012-08-25	FI_CLERK	3000.00	112	90

Figure 10.1 Employees Table

DEPT_ID	DEPT_NAME	MANAGER_ID	LOCATION_ID
10	Administration	200	1700
20	Marketing	201	1800
30	Purchasing	114	1700
40	Human Resources	203	2400
50	Shipping	121	1500
60	IT	103	1400
70	Public Relations	204	2700
80	Sales	145	2500
90	Accounting	205	1700
100	Customer Service	203	2400
11	Warehousing	114	2400

Figure 10.2 Departments Table

LOCATION_ID	STREET_ADDRESS	POSTAL_CODE	CITY	STATE_PROVINCE	COUNTRY_ID
1300	9450 Kamiya-cho	6823	Hiroshima		JP
1400	2014 Jabberwocky Rd	26192	Southlake	Texas	US
1500	2011 Interiors Blvd	99236	South San Francisco	California	US
1600	2007 Zagora St	50090	South Brunswick	New Jersey	US
1700	2004 Charade Rd	98199	Seattle	Washington	US
1800	147 Spadina Ave	M5V 2L7	Toronto	Ontario	CA
1900	6092 Boxwood St	YSW 9T2	Whitehorse	Yukon	CA
2000	40-5-12 Laogianggen	190518	Beijing		CN
2200	12-98 Victoria Street	2901	Sydney	New South Wales	AU
2400	8204 Arthur St		London		UK

Figure 10.3 Locations Table

Question 1: Write a query in SQL to display the employee names and department name for all employees in department 40.

Answer: SELECT E.first_name, E.last_name, D.dept_name
                FROM employees E
                JOIN departments D
                ON E.dept_id = 40 AND E.dept_id = D.dept_id;

Oracle SQL			T-SQL			MySQL		
FIRST_NAME	LAST_NAME	DEPT_NAME	first_name	last_name	dept_name	first_name	last_name	dept_name
Susan	Mavris	Human Resources	Susan	Mavris	Human Resources	Susan	Mavris	Human Resources

Figure 10.4 Query output for question 1

Using Older JOIN Style

                SELECT E.first_name, E.last_name,   D.dept_name
                FROM employees E, departments D
                Where E.dept_id = 40 AND E.dept_id = D.dept_id;

Oracle SQL			T-SQL			MySQL		
FIRST_NAME	LAST_NAME	DEPT_NAME	first_name	last_name	dept_name	first_name	last_name	dept_name
Susan	Mavris	Human Resources	Susan	Mavris	Human Resources	Susan	Mavris	Human Resources

Figure 10.5 Query output for question 1 (Old JOIN style)

Question 2: Write a query in SQL to display the full name of the employees and the department names.

Answer (Oracle):

> SELECT first_name || ' ' || last_name AS Full_Name, Dept_Name
> FROM employees E
> **JOIN** departments D
> **ON** (E.Dept_ID = D.Dept_id);

> SELECT first_name || ' ' || last_name AS Full_Name, Dept_Name
> FROM employees E
> **JOIN** departments D
> **USING** (Dept_ID);

Answer (T-SQL):

> SELECT first_name + ' ' + last_name AS Full_Name, Dept_Name
> FROM employees E
> **JOIN** departments D
> **ON** (E.Dept_ID = D.Dept_id);

Answer (MySQL):

> SELECT CONCAT(first_name, ' ', last_name) AS Full_Name, Dept_Name
> FROM employees E
> **JOIN** departments D
> **ON** (E.Dept_ID = D.Dept_id);

Oracle SQL		T-SQL		MySQL	
FULL_NAME	DEPT_NAME	Full_Name	Dept_Name	Full_Name	Dept_Name
Douglas Grant	Shipping	Douglas Grant	Shipping	Douglas Grant	Shipping
Adam Fripp	Shipping	Adam Fripp	Shipping	Adam Fripp	Shipping
Jennifer Whalen	Administration	Jennifer Whalen	Administration	Jennifer Whalen	Administration
Michael Hartstein	Marketing	Michael Hartstein	Marketing	Michael Hartstein	Marketing
Pat Fay	Marketing	Pat Fay	Marketing	Pat Fay	Marketing
Susan Mavris	Human Resources	Susan Mavris	Human Resources	Susan Mavris	Human Resources
Shelley Higgins	Sales	Shelley Higgins	Sales	Shelley Higgins	Sales
William Gietz	Sales	William Gietz	Sales	William Gietz	Sales
Steven King	Administration	Steven King	Administration	Steven King	Administration
Lex De Haan	Administration	Lex De Haan	Administration	Lex De Haan	Administration
Bruce Ernst	IT	Bruce Ernst	IT	Bruce Ernst	IT
Diana Lorentz	IT	Diana Lorentz	IT	Diana Lorentz	IT
Nancy Greenberg	Accounting	Nancy Greenberg	Accounting	Nancy Greenberg	Accounting
Daniel Faviet	Accounting	Daniel Faviet	Accounting	Daniel Faviet	Accounting

Figure 10.6 Query output for question 2

JOINING More Than Two Tables

Question 3: Write a query in SQL to display the full name of the employees who working in any department located in Seattle.

Answer (Oracle):

> SELECT first_name | | ' ' | | last_name AS Full_name, Dept_Name
> FROM employees E
> JOIN departments D
> ON E.Dept_ID = D.Dept_ID
> JOIN locations L
> ON (D.location_ID = L.location_id)
> WHERE city = 'Seattle';

🔲 FULL_NAME	🔲 DEPT_NAME
Lex De Haan	Administration
Steven King	Administration
Jennifer Whalen	Administration
Daniel Faviet	Accounting
Nancy Greenberg	Accounting

Figure 10.7 Query output for question 3 (Oracle)

Answer (T-SQL):

> SELECT first_name + ' ' + last_name AS Full_name, Dept_Name
> FROM employees E
> JOIN departments D
> ON E.Dept_ID = D.Dept_ID
> JOIN locations L
> ON (D.location_ID = L.location_id)
> WHERE city = 'Seattle';

Full_name	Dept_Name
Jennifer Whalen	Administration
Steven King	Administration
Lex De Haan	Administration
Nancy Greenberg	Accounting
Daniel Faviet	Accounting

Figure 10.8 Query output for question 3 (T-SQL)

Answer (MySQL):

> SELECT CONCAT(first_name, ' ', last_name) AS Full_Name, Dept_Name
> FROM employees E
> **JOIN** departments D
> **ON** E.Dept_ID = D.Dept_ID
> **JOIN** locations L
> **ON** (D.location_ID = L.location_id)
> WHERE city = 'Seattle';

Full_Name	Dept_Name
Jennifer Whalen	Administration
Steven King	Administration
Lex De Haan	Administration
Nancy Greenberg	Accounting
Daniel Faviet	Accounting

Figure 10.9 Query output for question 3 (MySQL)

LEFT JOIN

Suppose that we have the following order table and customer table:

ORDER_ID	ORDER_DATE	AMOUNT	CUSTOMER_ID
501	02-MAY-16	1000	2
502	06-DEC-15	300	3
503	20-JAN-16	500	2
504	05-FEB-17	2000	5
505	10-SEP-16	600	1

Figure 10.10 Order table

ID	FirstName	LastName	City	Country	Phone
1	Howard	Bell	Seattle	USA	(678)-555-7629
2	Alice	Carter	Southlake	USA	(617)-213-6874
3	Carine	Schmitt	Nantes	France	40.32.21.21
4	Paolo	Accorti	Torino	Italy	011-4988260
5	Helen	Bennett	Barcelona	Spain	(93)-203-4560

Figure 10.11 Customer table

From the order table you can see that customer No. 4 does not place any order.

Question 4: Write a query to display customer name, order amount and order date. List all the customer who have made or have not made orders.

Answer (Oracle & MySQL):

> SELECT C.ID, FirstName, O.Amount, O.Order_DATE
> FROM Customer C
> **LEFT JOIN** Orders O
> **ON** C.ID = O.Customer_ID;

ID	FIRSTNAME	AMOUNT	ORDER_DATE		ID	First Name	AMOUNT	Order_DATE
2	Alice	1000	02-MAY-16		2	Alice	1000.00	2016-05-02
3	Carine	300	06-DEC-15		2	Alice	500.00	2016-01-20
2	Alice	500	20-JAN-16		3	Carine	300.00	2015-12-06
5	Helen	2000	05-FEB-17		5	Helen	2000.00	2017-02-05
1	Howard	600	10-SEP-16		1	Howard	600.00	2016-09-10
4	Paolo	(null)	(null)		4	Paolo	NULL	NULL
100	Jason	(null)	(null)					

Figure 10.12 Query output for question 4 (Left: Oracle, Right: MySQL)

Answer (T-SQL):

> SELECT C.ID, FirstName, O.Amount, O.Order_DATE
> FROM Customer C
> **LEFT JOIN** Orders O
> ON C.ID = O.Customer_ID
> Order BY FirstName;

ID	FirstName	AMOUNT	Order_DATE
2	Alice	1000	2016-05-02
3	Carine	300	2015-12-06
2	Alice	500	2016-01-20
5	Helen	2000	2017-02-05
1	Howard	600	2016-09-10
4	Paolo	NULL	NULL

Figure 10.13 Query output for question 4 (T-SQL)

The first table (LEFT) is the Customer table. Left Join will display all the records from the Customer (LEFT) table. If no match (for customers who do not make orders) the result is NULL from the Order (RIGHT) table. In this case customer No. 4 Paolo's record has null value in Amount and Order_Date fields.

RIGHT JOIN and FULL JOIN are not used very often so we are not going to list samples here.

Recommended SQL Writing Style

1. Using upper case letters for SQL keywords.

For example, which style is easy to read for the following statements?

> Select c.id, firstname, o.amount, o.order_date
> from customer c
> left join orders o
> on c.id = o.customer_id order by firstname;

> SELECT C.ID, FirstName, O.Amount, O.Order_DATE
> FROM Customer C
> **LEFT JOIN** Orders O
> ON C.ID = O.Customer_ID Order BY FirstName;

2. Using multiple lines for longer statements.

For example, which style is easy to read for the following statements?

> Select c.id, firstname, o.amount, o.order_date from customer c left join orders o
> on c.id = o.customer_id order by firstname;

> SELECT C.ID, FirstName, O.Amount, O.Order_DATE
> FROM Customer C
> **LEFT JOIN** Orders O
> ON C.ID = O.Customer_ID Order BY FirstName;

3. Using a semicolon to end a statement.

Oracle requires a semicolon at the end of a statement. Although T-SQL and MySQL do not require a semicolon to end a statement, it's recommended to use semicolons at the end of statements.

4. Using comments.

Comment on a single line: /* Comments */
                                     -- Comments

Comment on multiple lines: /*
                                      * Comments
                                      * Comments
                                      */

Summary

Chapter 10 covers the following:

*   How to join tables using Inner Join.
*   How to use join with using clause in Oracle.
*   How to join more than two tables.
*   How to join tables using Left Join.

Exercises

10.1

Write a query to display number of employees in the department.

10.2

Write a query to display department name and city name.

10.3

Write a query to display employee name and country where he works.

Chapter 11
Views

A view is a virtual table. It looks like a table. A view is created in a SQL statement using one or more tables (views). Because views do not store data they only take small amount of disk space. Views can contains certain fields from a table. A Database administrator can create a view for non-sensitive data and set permission for the view to users.

Syntax

> **CREATE VIEW** view_name AS
> SELECT column(s)
> FROM tables
> [WHERE conditions];

Creating Views in Oracle

Since the Scott account does not have privilege to create a view, we will use the Oracle HR schema to create a view.

Step 1: Go to Command Prompt, type:

> C: \> sqlplus / as sysdba

At SQL prompt, type:

> Alter user hr IDENTIFIED by hr account unlock;

This command will unlock the HR account with password "hr".

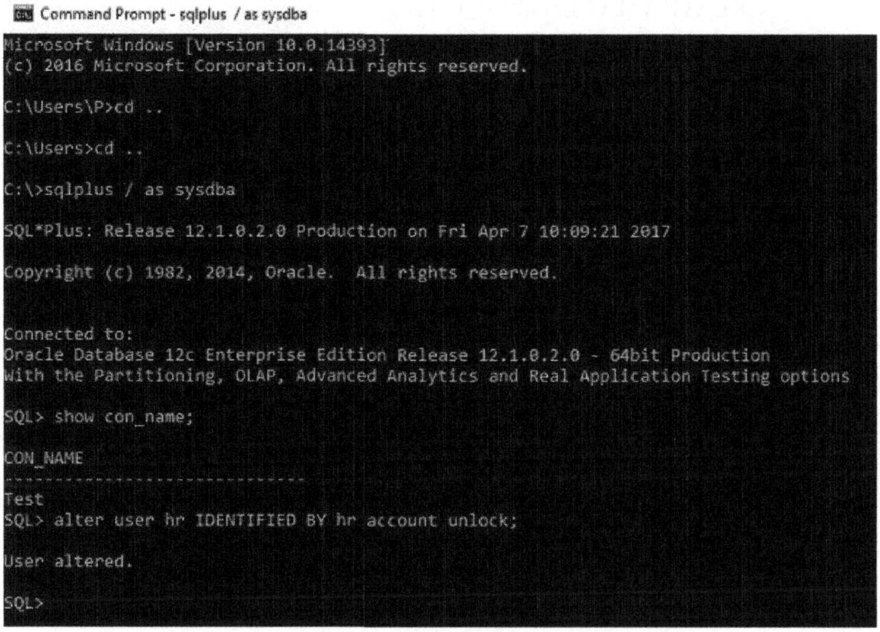

Figure 11.1 Unlock hr account in Oracle

Step 2: Connect the HR schema with username and the password:

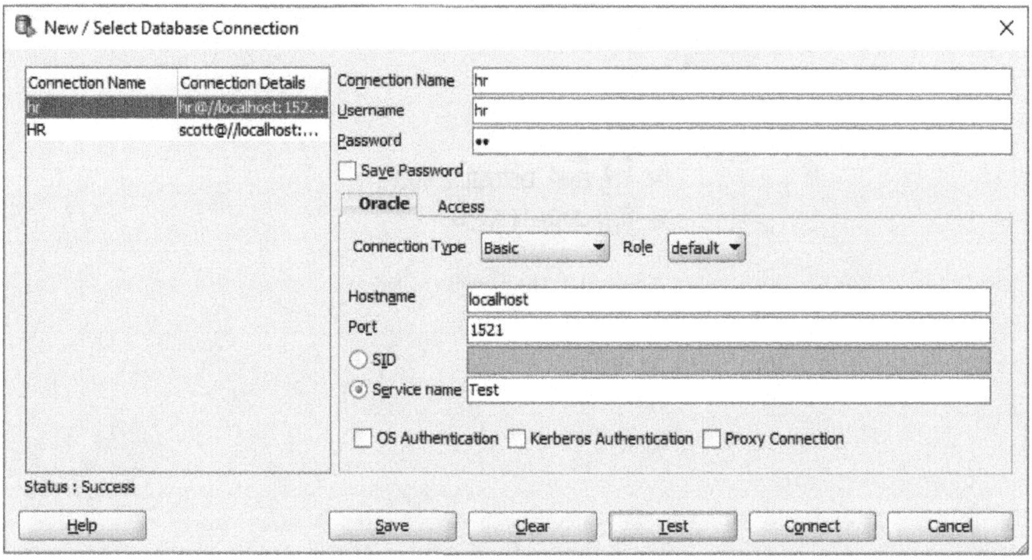

Figure 11.2 Oracle database connection

Step 3: Enter the Create View code on the SQL worksheet:

> **CREATE VIEW** Location_US **AS**
> Select Location_ID, City, State_Province
> From Locations
> Where Country_ID = 'US';

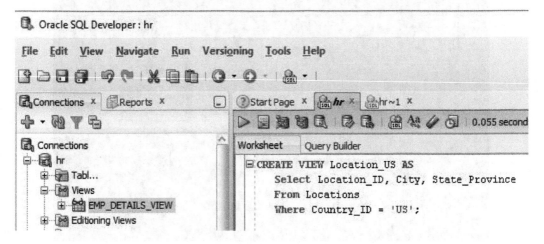

Figure 11.3 Creating a view

Step 4: Run the query to create the view then reconnect to HR schema:

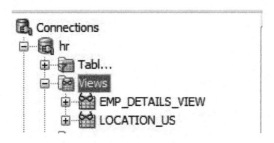

Figure 11.4 The view is created

Step 5: Check the result of the view:

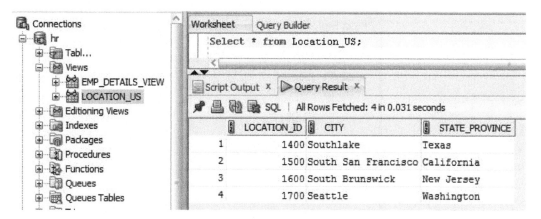

Figure 11.5 Executing the view

Creating Views in T-SQL

Step 1: Enter the Create View commands and run the query by click the **Execute** button.

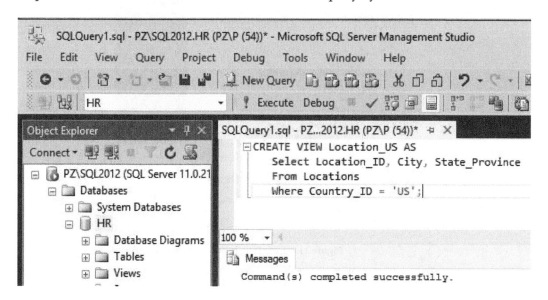

Figure 11.6 Creating a view

Step 2: Refresh the HR schema to see the view dbo.Location_US.

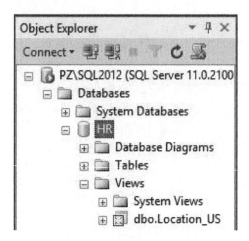

Figure 11.7 The view is created

Step 3: Right click the view and choose "**Select Top 1000 Rows**":

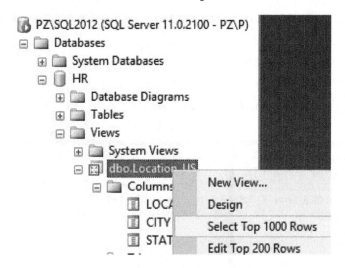

Figure 11.8 Executing a view

Step 4: Check the result of the view:

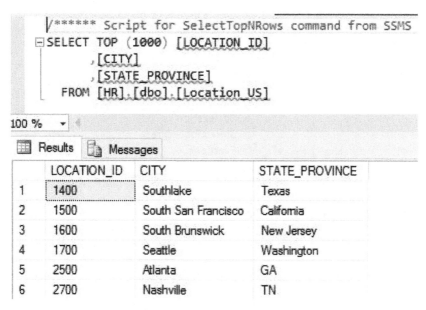

Figure 11.9 Output from the view

Creating Views in MySQL

Step 1: Enter the Create View commands and run the query by click the Execute button:

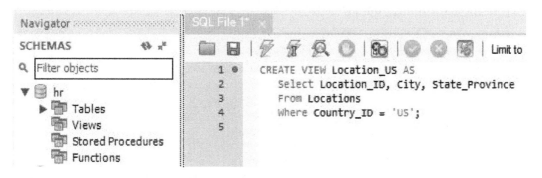

Figure 11.10 Creating a view

Step 2: Refresh the HR schema:

Figure 11.11 The view is created

Step 3: Clicking the icon next to "location_us":

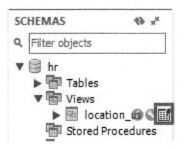

Figure 11.12 Executing the view

Step 4: Check the result from the view.

Location_ID	City	State_Province
1400	Southlake	Texas
1500	South San Francisco	California
1600	South Brunswick	New Jersey
1700	Seattle	Washington
2500	Atlanta	GA
2700	Nashville	TN

```
1 •    SELECT * FROM hr.location_us;
```

Figure 11.13 Output from the view

Updating Views

Syntax (Oracle)

> **CREATE OR REPLACE VIEW** view_name AS
> SELECT column(s)
> FROM tables
> [WHERE conditions];

Syntax (T-SQL & MySQL)

> **ALTER VIEW** view_name AS
> SELECT column(s)
> FROM tables
> [WHERE conditions];

Example (Oracle):

> **CREATE OR REPLACE VIEW**
> Location_US AS
> SELECT Location_ID, City, Country_ID
> FROM Locations
> WHERE Country_ID = 'US';

LOCATION_ID	CITY	COUNTRY_ID
1400	Southlake	US
1500	South San Francisco	US
1600	South Brunswick	US
1700	Seattle	US

Figure 11.14 Output for Oracle CREATE OR REPLACE VIEW (data from the original HR schema)

Example (T-SQL):

> **ALTER VIEW** Location_US AS
> SELECT Location_ID, City, Country_ID
> FROM Locations
> WHERE Country_ID = 'US';

Location_ID	City	Country_ID
1400	Southlake	US
1500	South San Francisco	US
1600	South Brunswick	US
1700	Seattle	US
2500	Atlanta	US
2700	Nashville	US

Figure 11.15 Output for T-SQL ALTER VIEW example

Example (MySQL):

> **ALTER VIEW** view_name AS
> SELECT columns
> FROM Locations
> WHERE conditions ID = 'US';

Location_ID	City	Country_ID
1400	Southlake	US
1500	South San Francisco	US
1600	South Brunswick	US
1700	Seattle	US
2500	Atlanta	US
2700	Nashville	US

Figure 11.16 Output for MySQL ALTER VIEW example

Deleting Views

Syntax (Oracle, T-SQL & MySQL)

> **DROP VIEW** view_name;

Example:

> **DROP VIEW** Location_US;

Summary

Chapter 11 covers the following:

- How to create a view in Oracle, T-SQL and MySQL
- How to update a view in Oracle, T-SQL and MySQL
- How to delete a view in Oracle, T-SQL and MySQL

Exercises

11.1

Create a view named v_employees to display the names and salary fields from the Employees table.

11.2

Drop the view.

Chapter 12

Data Import and Export

Data import and export are common tasks for developers or DBAs. Oracle, SQL Server and MySQL provide simple data import and export wizards. We will export data from the regions table then import the data using the exported csv file.

Oracle Data Export from Query Results

- First select data that you want to export. For example,
  SELECT * FROM regions;
- Right click the query result and select **Export...**

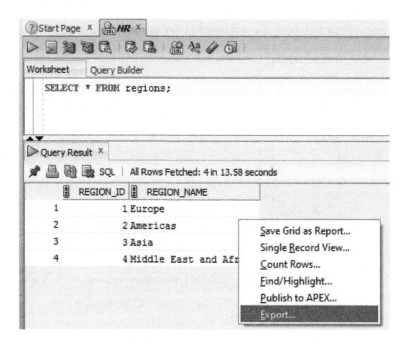

Figure 12.1 Exporting query result

- The Export Wizard step 1 of 2 screen shows up.

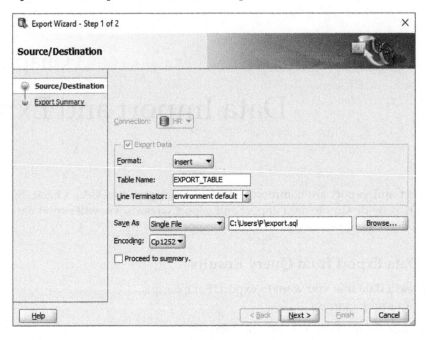

Figure 12.2 Export wizard

- In the Export Wizard change the Format to CSV and Encoding to UFT-8. Save the file as *regions.csv*.

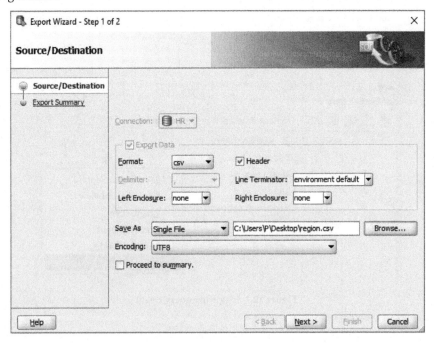

Figure 12.3 Export format

- Click Next to see the summary page.

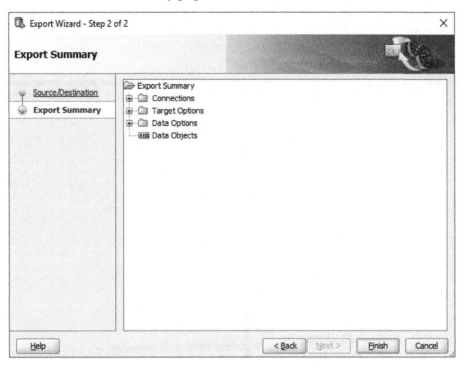

Figure 12.4 Summary

- Open the *regions.csv* to see the exported data.

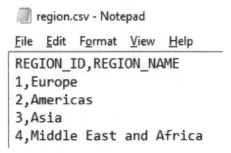

Figure 12.5 Checking the exported file

SQL Server Export Data from Query Results

- Let us select the data that we want to export in query worksheet:
    SELECT * FROM regions;
- Run the query and see the result.

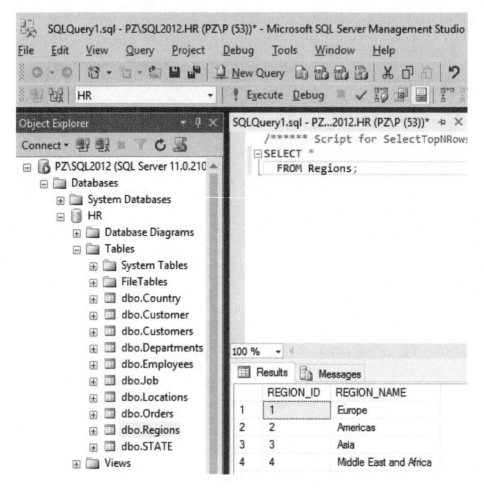

Figure 12.6 Running a query

- Right click the query result and select **Save Results As…**

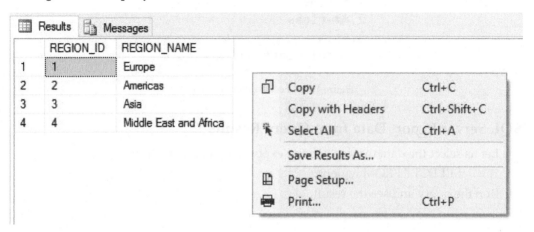

Figure 12.7 Saving query result

- Enter *region.csv* in the File name field.

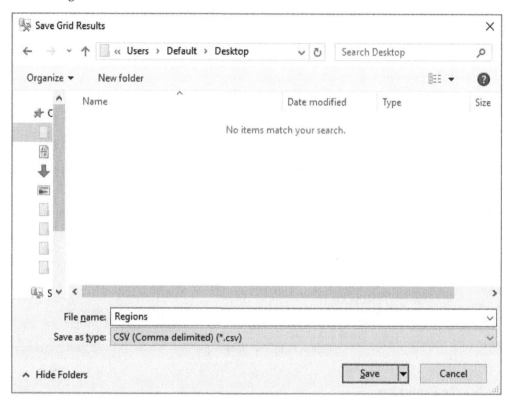

Figure 12.8 Export file name

- Open the *regions.csv* to see the exported data.

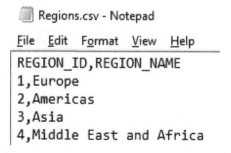

Figure 12.9 Checking the exported file

MySQL Export Data from Query Results

- Enter the following statement in query worksheet:
      SELECT * FROM regions;
- Run the query and see the output.

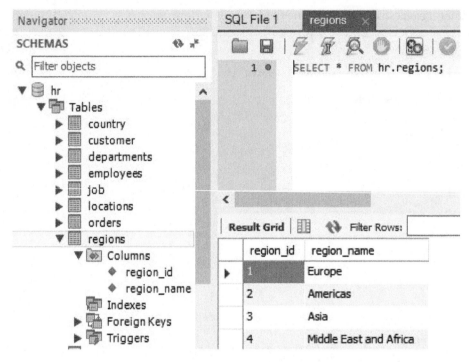

Figure 12.10 Executing a query

- Click the **Export** icon.

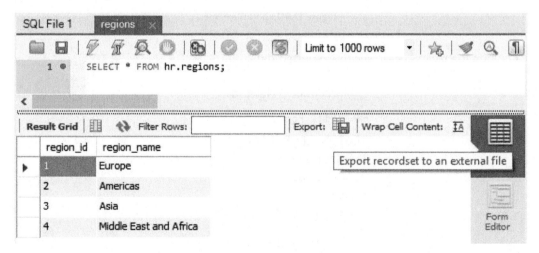

Figure 12.11 Export query result

- Enter *Regions.csv* in the File name field.

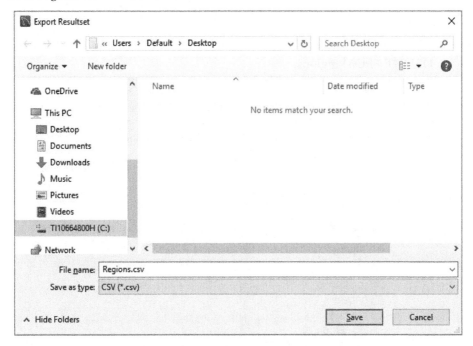

Figure 12.12 Entering the file name

- Open the *Regions.csv* to see the exported data.

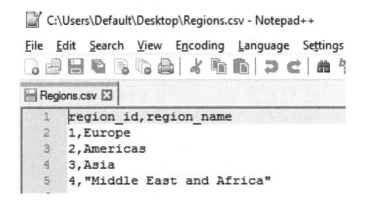

Figure 12.13 Checking the exported file

Oracle Data Import Tool

You have learned INSERT INTO statements in Chapter 6. It is used for inserting one or multiple records. For large amount of data we can use import wizard to insert data.

- Before inserting data we should prepare a table with corresponding data types in the data file. For demo purpose we just first delete records in the Regions table then import data from the regions.csv file.
- Enter the following statement:

  DELETE FROM regions;

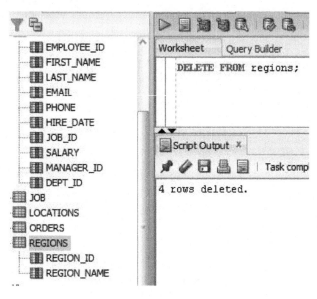

Figure 12.14 Deleting records in a table

- Then check the Regions table with the following statement:

  SELECT * FROM regions;

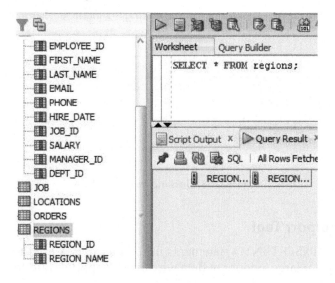

Figure 12.15 No data in Regions table after DELETE command

- Right click the Regions table and select **Import Data...**

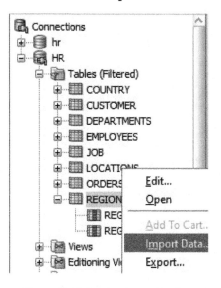

Figure 12.16 Starting importing data

- Select the *region.csv* file that we have exported.

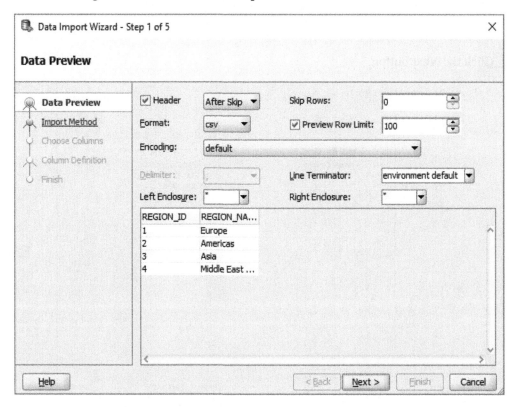

Figure 12.17 Opening the original file

- Follow the Data Import Wizard steps.

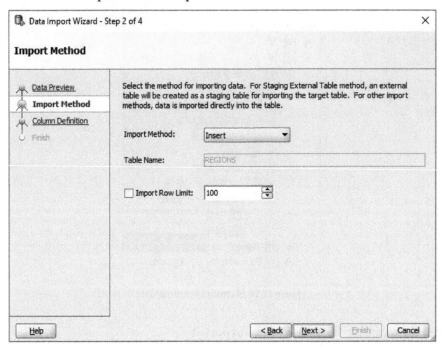

Figure 12.18　Selecting import method

- Click the **Next** button.

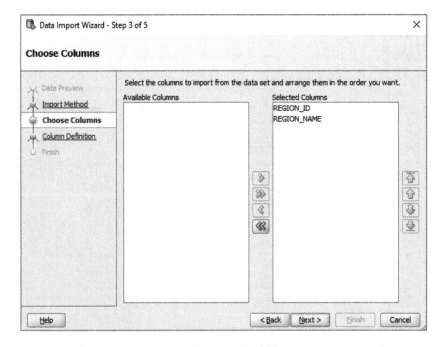

Figure 12.19　Columns in the file

- Follow the steps.

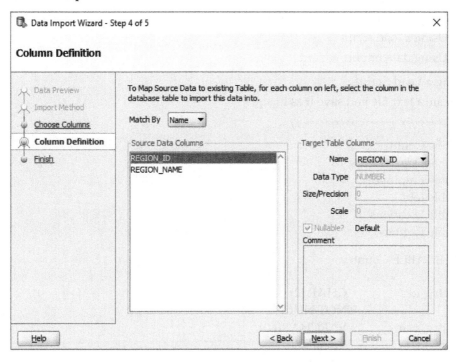

Figure 12.20 Comparing source and target data columns

- A message shows that the import data task is completed. Click **Finish** button on the next step.

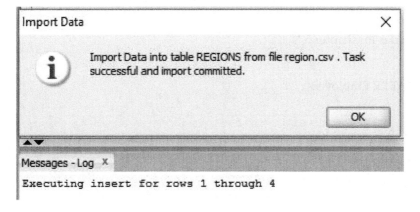

Figure 12.21 Import task is done

SQL Server Data Import Tools

Available tools to import data in SQL Server are:

1. Using a load script
2. Using data import wizard

1. Using a Load Script

- Create a text file and save it as hr.sql:

```
/* ****************************** */
drop table Country;
drop table Departments;
drop table Employees;
drop table Job;
drop table Locations;
drop table Regions;

CREATE TABLE Country
(
    country_id          CHAR (2)
                        NOT NULL,
    country_name        VARCHAR(40),
    region_id           smallint,
    PRIMARY KEY (country_ID),
    CONSTRAINT FK_RegCountry
    FOREIGNKEY (region_id)
REFERENCES Regions(Regins_ID)
);

CREATE TABLE Departments
(
...  /* see codes in Chapter 5 */
);

CREATE TABLE Employees
(
...
);

CREATE TABLE Job
(
...
);
```

```
CREATE TABLE Locations
(
...
);
```

```
CREATE TABLE Regions
(
...
);
```

```
INSERT INTO COUNTRY VALUES ('AR','Argentina',2);
... /* see codes in Chapter 6 */
```

```
INSERT INTO Employees
VALUES  (100,'Douglas','Grant','DGRANT','650.507.9844','23-Jan-08','SH_
CLERK',2600,114,50);
...
```

```
INSERT INTO DEPARTMENTS VALUES (10,'Administration',200,1700);
...
```

```
INSERT INTO Job VALUES ('AD_PRES','CEO',9000,20000);
...
```

```
INSERT INTO Locations
VALUES (1300,'9450 Kamiya-cho','6823','Hiroshima','','JP');
...
```

```
INSERT INTO REGIONS VALUES (1,'Europe');
...
/* ****************************** */
```

- Select the whole script and copy it to query worksheet. Run the script by clicking the **Execute** button.

2. Using data import wizard

- Enter the following statement and execute it:
    DELETE FROM regions;
- Check the Regions table with the following statement:
    SELECT * FROM regions;

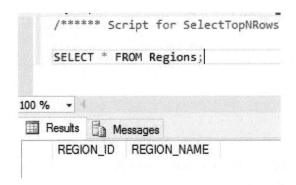

Figure 12.22 No data in the Regions table after DELETE command

- Right click a schema where you want to insert the data.
- Choose **Task -> Import Data…**

Figure 12.23 Starting import data

- In the Data source field select **Flat File Source** then select the **region.csv** file.

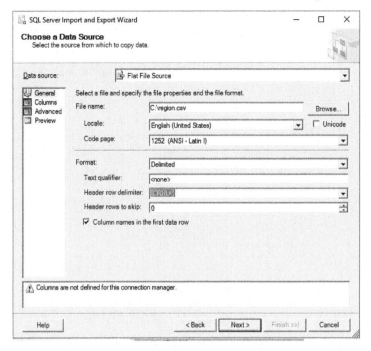

Figure 12.24 Selecting import file

- Follow the steps for SQL Server Import and Export Wizard.

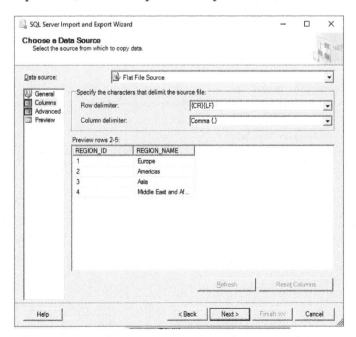

Figure 12.25 Choosing a data source

- Choose Destination as **SQL Server Native Client 11.0**

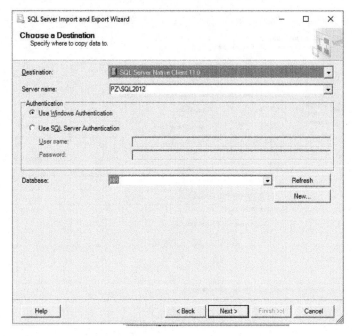

Figure 12.26 Choosing a destination

- Check the wizard summary page and click **Finish.**

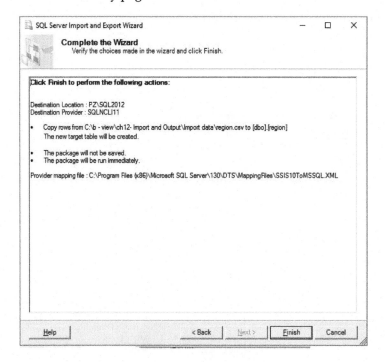

Figure 12.27 Summary page

- The data is imported successfully to Regions table.

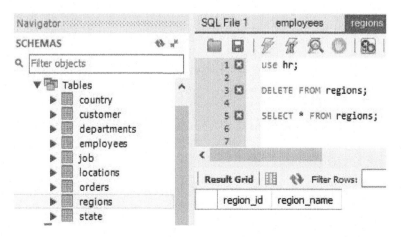

Figure 12.28 Importing successfully message

MySQL Data Import Tool

- Enter the following commands:
  use hr;
  DELETE FROM regions;
  SELECT * FROM regions;

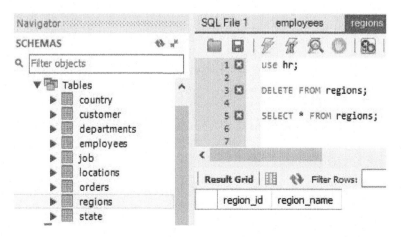

Figure 12.29 No data in Regions table after DELETE command

- Right click the regions table and choose **Table Data Import Wizard.**

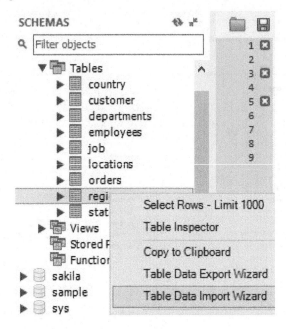

Figure 12.30 Starting data import

- Select *Regions.csv* in the File Path field.

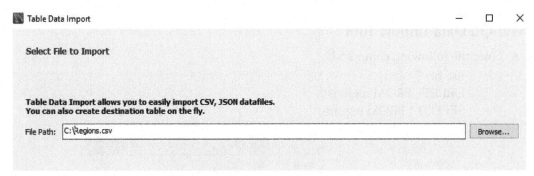

Figure 12.31 File path for the csv file

- You can choose existing table or create a new table.

Figure 12.32 Selecting destination

- The wizard will set the Encoding UTF-8 and match source and destination columns.

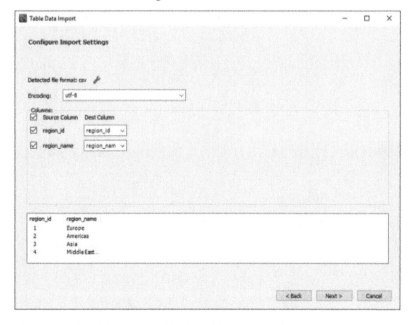

Figure 12.33 Source and destination columns

Summary

Chapter 12 covers the following:

- Exporting Oracle data from a query result
- Exporting SQL Server data from a query result
- Exporting MySQL data from a query result.
- How to import data to an Oracle table.
- How to import data to a SQL Server table.
- Using a load script for SQL Server data import.
- How to import data to a MySQL table.

Exercise

12.1

Export records in Employees table to a csv file. Create a new table or delete the records in the Employees table. Using the csv file to import data to the table.

Chapter 13

Stored Procedures

What is a Stored Procedure

When you create a useful query in your working place it's very possible that you need to run that query again. For example, you may need to run a monthly sales report automatically at the beginning of a month. Stored procedures can do that job for you. You can save the query in a stored procedure and schedule a task to run the job automatically.

A stored procedure usually has three parts:

- Declaration
- Execution
- Exception (Optional)

A Simple Stored Procedure

Syntax:

CREATE [OR REPLACE] PROCEDURE proc_name [(parameter1, parameter2 ...)]
IS | AS (Oracle)
AS (SQL Server)
    [declaration part]

BEGIN
    executable part

[EXCEPTION]
    exception part
END;

Let us create a stored procedure to count row numbers in the regions table. There is no input parameters and output values.

Steps to Create an Oracle Stored Procedure

- Enter SQL code in the query editor.
- Run the code to create a stored procedure.
- Enter **EXEC** proc_name to run the stored procedure.

- To display message you need to **SET SERVEROUPUT ON** and using **dbms_output.put_line().**

Table 13.1 Comparison for a simple stored procedure

Oracle PL/SQL	SQL SERVER T-SQL	MySQL

SET SERVEROUTPUT ON;
 	
CREATE or REPLACE
procedure p1
IS
  num INT;
BEGIN
 SELECT count(*) INTO num
 FROM regions;
dbms_output.put_line(num);
END;
 	
EXEC p1;
	CREATE procedure p1
	AS
	BEGIN
	 DECLARE @num INT;
	
	 SET NOCOUNT ON;
	 SELECT @num = count(*)
	 FROM regions;
	 PRINT @num;
	END;
	
	EXEC p1;
		DELIMITER //
		
		CREATE procedure p1()
		BEGIN
		 SELECT count(*)
		 FROM regions;
		END;
		//
		
		CALL p1();

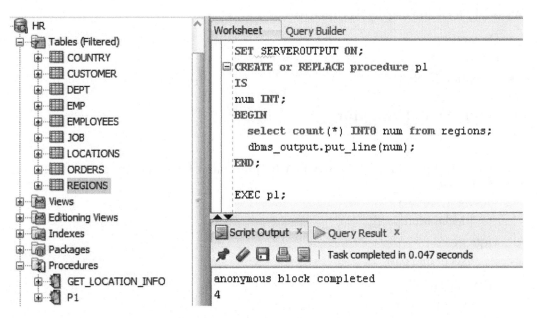

Figure 13.1 A simple Oracle procedure

Steps to Create a SQL Server Stored Procedure

- Enter SQL code in the query editor.
- Run the code to create a stored procedure.
- Delare valable(s) under **BEGIN** keyword.
- Enter **EXEC** proc_name to run the stored procedure.
- Using **SET NOCOUNT ON.**

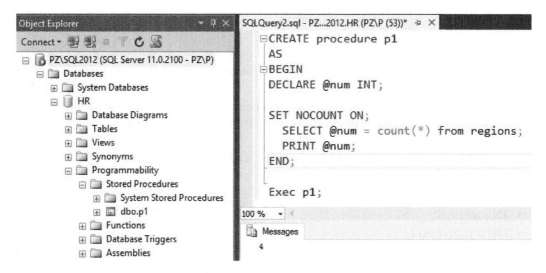

Figure 13.2 A simple SQL Server procedure

Steps to Create a MySQL Stored Procedure

- Create a delimiter like // or $$. The delimiter is characters that is used to complete an SQL statement.
- Enter SQL code in the query editor.
- Run the code to create a stored procedure.
- Enter **CALL** proc_name() to run the stored procedure.

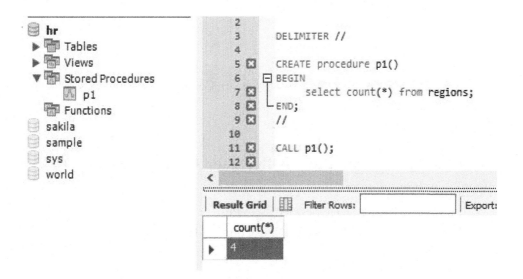

Figure 13.3 A simple MySQL procedure

A Stored Procedure with Parameters

Oracle Parameters

- **IN** (optional)—To pass value(s) to a stored procedure. The values are not changed in the procedure. IN keyword is optional.
- **OUT**—To get value(s) from a stored procedure. The value(s) can be passed to OUT parameter(s) inside the stored procedure. A calling program is needed to get the output value(s).
- **IN OUT**—To pass and get value(s) from a stored procedure.

SQL Server Stored Procedure Parameters

- **IN** (optional)
- **OUT | OUTPUT**

MySQL Server Stored Procedure Parameters

- **IN** (optional)
- **OUT**
- **INOUT**

To Create an Oracle Stored Procedure with IN and OUT Parameters:

```
CREATE or REPLACE procedure get_Location_Info
    (L_ID IN NUMBER,
     L_City OUT VARCHAR2,
     L_Country_ID OUT CHAR
    )
AS
BEGIN
    SELECT City, Country_ID INTO L_City, L_Country_ID FROM LOCATIONS
    WHERE LOCATION_ID = L_ID;
END get_Location_Info;
```

To Execute an Oracle Procedure with IN and OUT Parameters:

```
DECLARE
    Location_City LOCATIONS.CITY%TYPE;
    Location_Country_ID LOCATIONS.Country_ID%TYPE;
```

```
BEGIN
  get_Location_Info ( 1700, Location_City, Location_Country_ID );
  DBMS_OUTPUT.PUT_LINE ( Location_City || ' '|| Location_Country_ID);
END;
```

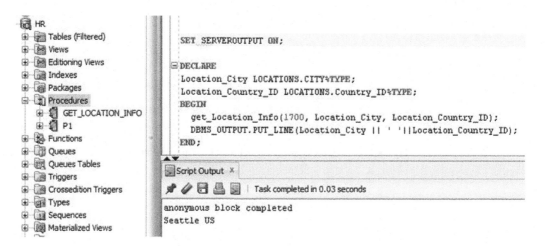

Figure 13.4 Calling Oracle procedure with IN and OUT parameters

To Create an Oracle Stored Procedure with IN OUT Parameters:

```
create or replace
procedure example_INOUT ( x IN OUT NUMBER)
AS
BEGIN
 x := x + 6;
END example_INOUT;
```

To Execute an Oracle Procedure with IN OUT Parameters:

```
DECLARE
  x number;
BEGIN
  x:= 10;
  example_INOUT ( x );
 DBMS_OUTPUT.PUT_LINE ( 'x is' || x);
END;
```

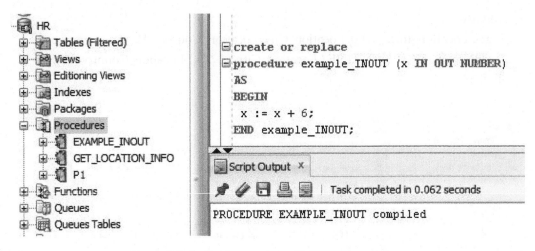

Figure 13.5 An Oracle procedure with IN OUT parameters

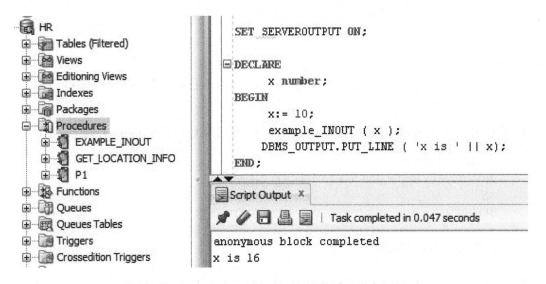

Figure 13.6 Calling Oracle procedure with IN OUT parameters

To Create a SQL Server Stored Procedure with IN Parameters:

```
CREATE procedure get_Location_Info
    (@L_ID FLOAT)
AS
BEGIN

 SET NOCOUNT ON;
    SELECT City, Country_ID FROM LOCATIONS
    WHERE LOCATION_ID = @L_ID;
END;
```

To Execute SQL Server Procedures with IN Parameters:

EXEC get_Location_Info 1700;

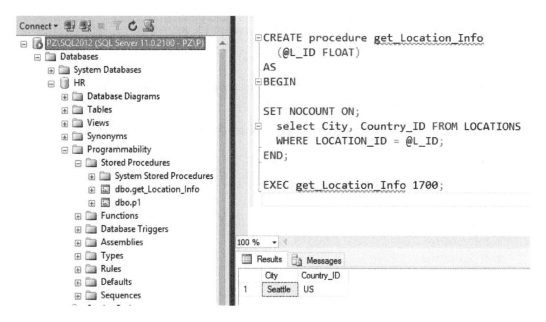

Figure 13.7 A SQL Server procedure with parameters

To Create a MySQL Stored Procedure with IN and OUT Parameters:

DELIMITER //

CREATE procedure get_Location_Info
    (@L_ID FLOAT,
    @L_CITY VARCHAR OUT,
    @L_Country_ID CHAR OUT)

AS
BEGIN
    SELECT @L_City =City, @L_Country_ID = Country_ID
    FROM LOCATIONS
    WHERE LOCATION_ID = @L_ID;
END;
    //

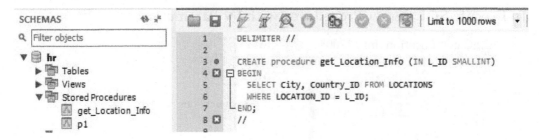

Figure 13.8 A MySQL procedure with IN and OUT parameters

To Execute a MySQL Procedure with IN and OUT Parameters:

CALL get_Lo cation_Info (1700);

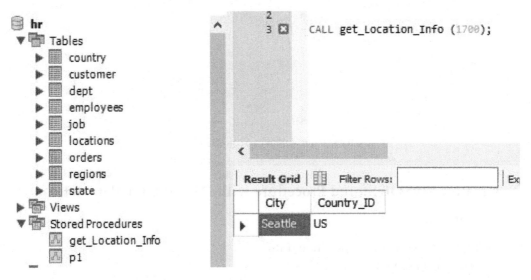

Figure 13.9 Calling a MySQL procedure with IN and OUT parameters

Summary

Chapter 13 covers the following:

- Basic structures of a stored procedure
- Steps to create a stored procedure in Oracle, SQL Server and MySQL
- A simple procedure without parameters
- Steps to Create a stored procedure with parameters in Oracle, SQL Server and MySQL

- A sample stored procedure with IN and OUT parameters
- A sample stored procedure with IN OUT parameters

Exercise

13.1

Create a stored procedure to list employees who work for shipping department.

Index

A

ADD_MONTH ( ) 113, 114
addition (+) 74
aggregate functions 89, 94, 97, 98
aliases 84
ALTER COLUMN 61, 62, 64
ALTER TABLE 60–64, 135
AND 77
ANSI 2
arithmetic operators 74, 76, 87
ASC (ascending order) 72
AVG ( ) 89, 90, 94, 97, 98

B

BETWEEN 8, 9, 12, 79, 88, 113, 116, 124, 125, 131, 137
BIGINT data type 12
BINARY data type 11
BLOB data type 11

C

CASE 9, 88, 108, 112, 122, 130–132, 136, 143, 144
CAST ( ) 117
CEILING ( ) 98
char data type 11, 12, 14
character data type 11, 14
CHECK constraint 7
CLOB data type 11
columns 4–7, 11, 14, 42, 54–56, 61–64, 67, 72–74, 77–79, 81, 82, 84, 88–92, 94, 95, 101, 123, 124, 126, 129, 146, 153, 154, 164, 165, 173
  column aliases 84
command line 34, 36, 40, 46
comments 144
comparison operator 76, 87
CONCAT ( ) 103–105
constraints 7, 52, 53, 166
conversion 117, 119, 120
CONVERT ( ) 117, 119
COUNT ( ) 89, 90, 95, 97, 98
CREATE DATABASE 46–48, 64, 87
CREATE TABLE 51–53, 55–57, 64, 134, 135, 166, 167
CREATE USER 48
CREATE VIEW 146, 148, 149, 151

D

Data Control Language (DCL) 47, 87
Data Definition Language (DDL) 47, 48, 65
Data Manipulation Language (DML) 47, 65, 87
data types 4, 11–14, 54–56, 117, 119, 123, 162
database administrators (DBAs) 1
date and time data type 13, 14
DATE_ADD ( ) 113, 114
DATE_DIFF ( ) 113, 116
DECIMAL 8, 12, 13, 53, 62, 63, 90, 101, 102, 106, 121
DELETE 4, 7, 47, 51, 60, 65, 77, 78, 86–88, 154, 162, 167, 168, 171, 174
DESC 72, 82, 83, 94
DISTINCT 73, 123, 125, 126
division (/) 74
DOUBLE data type 12
DROP TABLE 51, 60, 166
DROP VIEW 154

E

ENUM data type 11
Entity Relationship Diagram (ERD) 8, 10
Equals operator (=) 76
expression 74, 78–82, 98, 117, 119, 131
EXTRACT 111, 113–115

F

fields 5, 7, 12, 74, 109, 111, 114, 132, 143, 146, 154, 159, 161, 169, 172
fixed length 11, 12
FLOAT data type 4, 12
FLOOR ( ) 98, 99
foreign keys 6–9, 52, 53
FORMAT ( ) 103, 106
functions 84, 89–98, 101, 103, 105–120

G

GETDATE ( ) 113, 115, 119,
GRANT 4, 47, 48, 65, 66, 87, 88, 167
greater than operator (>) 76
GROUP BY 89, 94–97

H

HAVING clause 94, 97
history 1, 2, 10

I

IMAGE data type 11
IN condition 78
INNER JOIN 137, 144
INSERT INTO 47, 65–71, 85, 87, 133–135, 161, 167
installation 15, 16, 19–21, 26, 30, 32, 38, 40, 42, 43, 46, 49
INT data type 12
integrity 7
    Entity Integrity 7
    Referential Integrity 7
IS NULL 79, 80, 137, 143
IS NOT NULL 80

J

JOIN 137–144
    INNER JOIN 137, 144
    LEFT JOIN 137, 142–144
    RIGHT JOIN 137, 143
    FULL JOIN 137, 143

K

keys 9
    primary keys 9
    foreign keys 9

L

languages 1, 2, 4, 11, 17, 47, 65, 87
LEFT JOIN 137, 142–144
LEN ( ) 103, 107
LENGTH ( ) 103, 107
less than operator (<) 76
LIKE condition 81
LIMIT 122, 126, 128, 136, 176, 177, 181
LONGBLOB data type 11
LONGTEXT data type 11
LOWER ( ) 103, 108
LTRIM ( ) 103, 109

M

master database 42, 43
MAX ( ) 90, 92, 97, 98
MEDIUMTEXT data type 11
Microsoft SQL Server 2
MIN ( ) 90, 91, 97, 98
Minutes 114, 120
MONEY data type 12
Month 113, 114, 116, 120, 175
MONTH_BETWEEN ( ) 116
multiplication (*) 74
MySQL Server 31, 37, 178

N

NCHAR data type 11
NOT IN 78, 122, 125, 126, 136
NOT NULL 7, 14, 52, 53, 80, 134, 135, 166
NTEXT 11
NULL 7, 14, 52, 53, 79, 80, 134, 135, 137, 143, 166
NUMBER data type 12, 14
numeric data type 12, 13
NVARCHAR data type 11

O

ON 87, 137, 139–144, 176, 180
OR 77
ORDER BY 72, 82, 83, 94, 95, 123, 124, 143, 144
operating system 15

P

PERIOD_DIFF ( ) 113, 116
POWER ( ) 101
primary key 6–9, 52, 53, 132, 135, 166
privileges 47, 48, 87, 146
    GRANT 47, 48, 87
    REVOKE 47, 87

Q

query worksheet 40, 48, 66, 157, 159, 167

R

RAW data type 11
records 6, 7, 9, 47, 61, 65, 72, 77, 82, 85, 86, 90–92, 126, 132, 133, 137, 143, 161, 162, 174
REAL data type 12
REFERENCES 52, 53, 166
referential integrity 7
relational database 1–3, 10, 15, 137
revoking privileges 87
RIGHT ( ) 103, 110
ROUND ( ) 90, 97, 98, 101
ROWNUM 122, 126, 127, 136
rows 6–9, 47, 51, 60, 61, 74, 89, 105, 108, 110, 112, 123–127, 137, 150, 175
RTRIM ( ) 103, 111

S

schema 8, 35, 39, 45, 49, 50, 51, 146–148, 150, 152, 153, 168
SELECT 47, 56, 57, 65, 72–87, 89, 90–96, 99–103, 105–120, 123–131, 134, 135, 137–143, 144, 146, 153–155, 157, 159, 162, 167, 171, 176, 178, 180, 181
SET data type 11
SMALLDATETIME 13
SMALLINT data type 12
SMALLMONEY data type 12
SQL Server Management Studio 26, 34, 40–42, 46

SQRT ( ) 98, 102
Stored procedure 175–183
strings 4, 11, 12, 103, 104, 106–112, 118, 120, 121
Structured Query Language 1, 2
subqueries 128
SUBSTR 103, 111, 112
SUBSTRING 103, 111, 112
subtraction (–) 74
SUM ( ) 90, 93, 97, 98

T

Table aliases 84
Tables 2, 4–9, 11, 12, 14, 35, 42, 46–48, 51–74, 77–79, 81,
    82, 84–87, 89–94, 98, 104, 118, 122–128, 133–139,
    141–144, 146, 153–155, 162, 166–168, 171–175
   creating 4, 9, 14, 47, 48, 51–55, 64, 65, 146
   DROP 4, 47, 51, 60, 64, 154, 166
   joins 137–139, 141–144
TEXT data type 11
TINYINT data type 12
TINYTEXT data type 11
TOP 45, 105, 108, 110, 112, 122, 126, 128, 136, 150
Transact-SQL (T-SQL) 1

Truncate Table 2, 61, 64
TRUNC ( ) 98, 102

U

Unicode 36
Union 123–124, 136
Union ALL 123–124, 136
UNIQUE 7, 124
UPDATE 1, 4, 47, 65, 77, 78, 86, 87, 154

V

VARBINARY data type 11
VARCHAR data type 11, 12
VARCHAR2 data type 11, 12
Views 46, 146, 148–154
   creating 146, 148, 149, 151
   update 154
   deleting 154

W

Where clause 74, 127, 128
Wildcards 81

About the Author

Preston Zhang has over 20 years of experiences in database design and implementation. As a database administrator, he manages Oracle, SQL Server and MySQL database servers for university departments in Georgia. He has written many queries in Oracle SQL, SQL Server T-SQL and MySQL to process millions of records for business reports. He has developed Web applications using Oracle database as back-end for a large health care company. He has taught undergraduate database and programming courses in private universities for over 10 years. He has a Master of Science degree in Computer Information Systems from University of Wisconsin-Parkside. He lives in Georgia with his family and can be reached at prestonz668@gmail.com.